D1289493

GAS CHROMATOGRAPHY IN FORENSIC SCIENCE

ELLIS HORWOOD SERIES IN FORENSIC SCIENCE
Series Editor: Dr JAMES ROBERTSON, Head of Forensic Services Division, Australian Federal Police, Canberra, Australia

Aitken and Stoney	**THE USE OF STATISTICS IN FORENSIC SCIENCE**
Cole and Caddy	**THE ANALYSIS OF DRUGS OF ABUSE: An Instruction Manual**
Ellen	**THE SCIENTIFIC EXAMINATION OF DOCUMENTS**
Pounder	**SUDDEN NATURAL DEATH: Forensic Pathology and Medico-Legal Investigation**
Robertson, Ross and Burgoyne	**DNA IN FORENSIC SCIENCE: Theory, Techniques and Applications**
Robertson	**FORENSIC EXAMINATION OF FIBRES**
Tebbett	**GAS CHROMATOGRAPHY IN FORENSIC SCIENCE**

GAS CHROMATOGRAPHY IN FORENSIC SCIENCE

Editor I. TEBBETT
Department of Pharmacodynamics, University of Illinois at Chicago, USA

ELLIS HORWOOD
NEW YORK LONDON TORONTO SYDNEY TOKYO SINGAPORE

First published in 1992 by
ELLIS HORWOOD LIMITED
Market Cross House, Cooper Street,
Chichester, West Sussex, PO19 1EB, England

A division of
Simon & Schuster International Group
A Paramount Communications Company

Printed and bound in Great Britain
by Bookcraft, Midsomer Norton

British Library Cataloguing in Publication Data

A catalogue record for this book is available from the British Library

ISBN 0–13–327198–6

Library of Congress Cataloging-in-Publication Data

Available from the publisher

Table of contents

Preface

Gas chromatography was introduced into analytical chemistry in the early 1950s. By 1960, after a slow start, increasingly large numbers of papers describing gas chromatography applications to analytical chemistry problems began to appear in the literature. This was associated with the introduction to the market of several lines of commercially produced instruments. Since these early beginnings, gas chromatography has developed into an extremely sensitive technique, capable of dealing with small amounts of analyte contained within complex mixtures.

Although it has the limitation that the species to be analyzed must be capable of being volatilized, gas chromatography is directly applicable to the analysis of many forensic samples. Gas chromatography used alone or in combination with a mass spectrometer has thus become invaluable to the forensic laboratory. Applications have been described for the analysis of blood alcohol, drugs of abuse, either as street drugs or as toxicological samples in biological fluids, debris from fire scenes suspected of containing accelerants, explosives and even trace evidence such as paints, polymers and plastics. Gas chromatography is especially useful for the examination of complex mixtures of volatile organic substances. For such samples this technique has no peer.

While there are many books which discuss the theory and applications of gas chromatography, none have dealt solely with analytical problems peculiar to the examination of forensic samples. We have attempted to address these problems by drawing upon the expertise of individuals working in very different areas of forensic science but brought together by their use of this one technique.

1

Forensic gas chromatography

David T. Staffford, Ph.D.
University of Tennessee, Toxicology Laboratory, 3 N. Dunlap, Memphis, TN 38163, USA

1.1 INTRODUCTION

In attempting to address the solution of most problems of any degree of complexity, it is frequently necessary to divide the problem into a series of parts which may be resolved in a manner which will elucidate the approach to solution of the primary problem. Forensic analyses almost invariably require this type of treatment. In physical/chemical analyses of mixtures or solutions the analyst is most fortunate to have available to him some rather powerful tools to pursue these analyses. If the material to be characterized is soluble in aqueous or organic solvents under reasonable conditions, i.e., at temperatures below 100°C, or having a vapor pressure of about 1 torr at 300°C or lower, with structural integrity, then chromatographic means of separation and characterization are usually appropriate. Liquid or supercritical fluid chromatography are techniques which may be considered for those compounds which do not have sufficient vapor pressure or thermal stability to be separated by gas chromatography (GC). It should be noted that these are three complementary techniques; no one of them is applicable to some analyses. This discussion will cover gas chromatography.

The use of chromatography was first reported by Ramsey (1905) and Tswett (1906); however, the process was tedious and was used primarily for separation of compounds derived from natural products and preparation of small amounts of their components for further study. Martin & Synge (1941) published their work on the development of the concept of partitioning. They subsequently received the Nobel Prize for this work. Gas chromatography was an outgrowth of developments by James & Martin (1951, 1952) and was the basis for the first commercial gas chromatograph in 1953. These first GCs were equipped with poorly thermostatted, large diameter, $\frac{1}{4}$ inch or larger, glass or metal columns packed with crushed firebrick coated with a variety of stationary phases from vacuum grease to detergents. The most important early detector was the thermal conductivity (TC) cell.

Recognition of the tremendous separating capability of GC promoted development of GC theory as well as instrumentation. By the late 1950s pioneers such as Van Deemter *et al.* (1956), Gluechauf (1955), and Keulemans (1957) had published much of the early GC theory; Golay (1957, 1958) had investigated and demonstrated wall coated open tubular (WCOT) or capillary column GC; and Pretorius (1979) and McWilliam & Dewar (1958) had described improvements in the flame ionization detector (FID).

Packed column GC was the standard for the first twenty years; however in the mid-1970s glass capillary columns and instrumentation designed for WCOT column chromatography began to be commercially available. Dandeneau & Zerenner (1979, 1990) introduced the use of fused silica capillary columns, and WCOT column GC has now largely replaced packed column GC except for the separation of fixed gases and preparative work.

Because of the great separating power it possesses, its relative simplicity of use, and its ability to interface readily with molecular identification methods such as infra-red spectroscopy and mass spectrometry, GC is usually the separation method of choice if the materials of interest have structural stability in the vapour phase, and a vapour pressure of at least one torr at 300°C or less. An estimated 18–20 percent of known organic compounds fit these criteria.

1.2 CHROMATOGRAPHY

Definition

Chromatography is a separation process which depends upon the differential distribution of sample components between a moving and a stationary phase. This definition covers all types of chromatography — gas, liquid, and supercritical fluid — with the exception of size exclusion chromatography. Please note that it does not indicate that chromatography is an identification technique — it is not; its power is in separation, its weakness is identification.

Differential distribution

If an analyte is placed in a system which consists of two immiscible phases it will distribute between these two phases in some manner. At equilibrium this distribution will be dependent upon the nature of the two phases, the analyte, and the system pressure and temperature. In gas chromatography a gas is chosen as one phase, the mobile phase, and a low volatility liquid is selected as the second phase, the stationary phase. Following the work of Nernst in 1891 a distribution coefficient, K_D, can be defined for chromatographic purposes:

$$K_D = \frac{\text{Concentration of component in the stationary phase}}{\text{Concentration of component in the mobile phase}}$$

Since GC operates at relatively low pressure, 20–25 psig usually, and because K_D is rather insensitive to the type of carrier gas, nitrogen, helium, or hydrogen, then the influence of these factors is small enough to be ignored. The distribution coefficient for a particular component for gas chromatographic purposes can be considered to be a function of only the stationary phase and operating temperature.

Each component in a sample will exhibit its own K_D, and as the concentrations in the GC column are quite low, K_D for each component is independent of the presence of other compounds in the sample. At a particular temperature and carrier gas flow rate the retention time of a component will be a function of the time spent in the stationary phase. A compound with a lower K_D will spend less time in the stationary phase than one with a higher K_D, and since each spends the same amount of time in the mobile phase, the one with the lower K_D will elute more quickly. In order for two components to be separated they must have different K_Ds, i.e., there must be differential distribution.

Resolution

The only reason to do chromatography is to separate or resolve one compound from another. Fig. 1.1 is a typical chromatogram of the separation of two components of

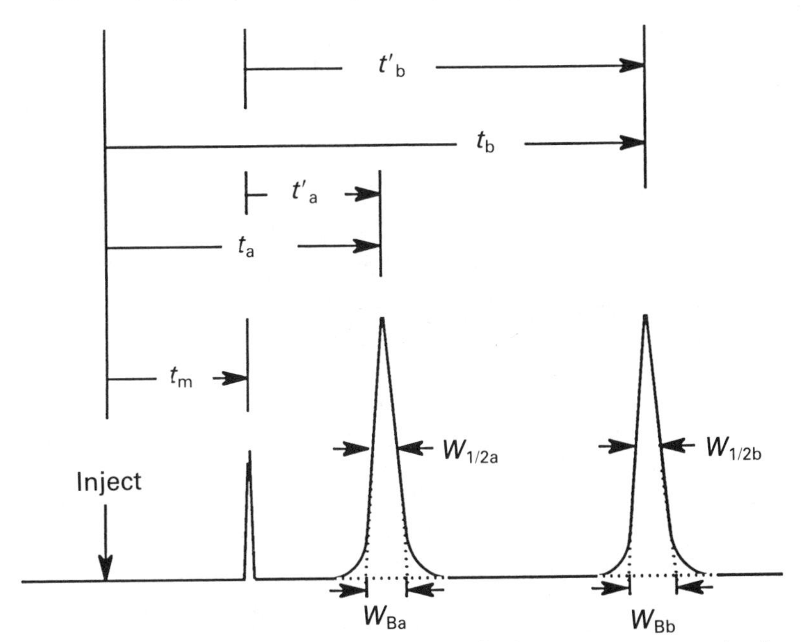

Fig. 1.1. Typical chromatogram. t_m = time for an unretained component. t_a = retention time for component 'a'. t'_a = adjusted retention time for component 'a'. W_{Ba} = width at base for peak 'a'. $W_{1/2a}$ = width at half-height for peak 'a'.

a sample, with some of the important features shown. In chromatographic practice resolution, R, is defined as:

$$R = \frac{(t_b - t_a)}{1/2(W_{Ba} + W_{Bb})}$$

For gaussian peaks a resolution of one, $R = 1$, the peak maxima will be separated by four times the average standard deviation about the two means. For $R = 1$ the separation of the two components will be approximately 98%; at $R = 1.5$, essentially

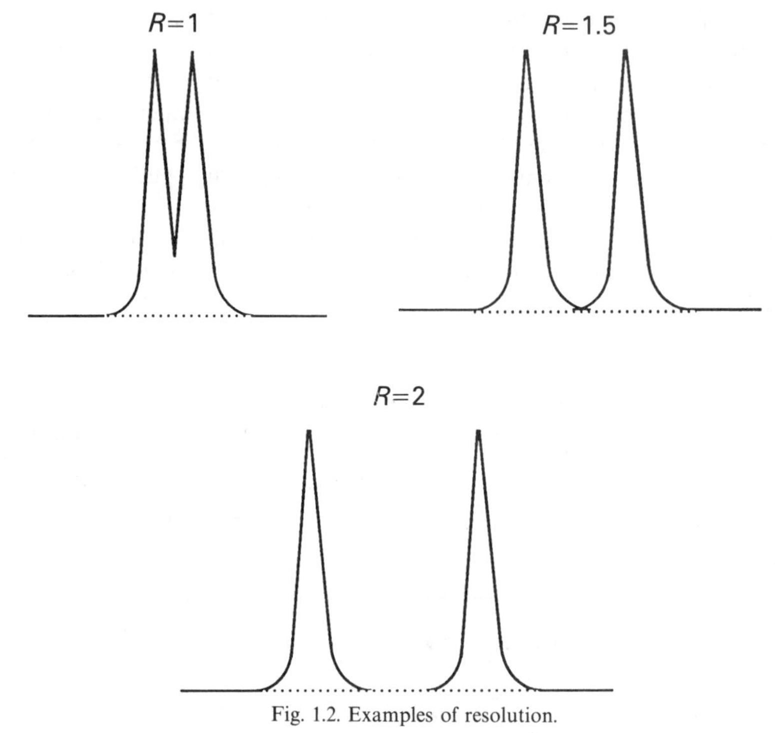

Fig. 1.2. Examples of resolution.

baseline resolution, the separation is approximately 99.7%. Fig. 1.2 illustrates several degrees of resolution.

The desirable, or acceptable resolution will depend upon the particular task to be accomplished. If the purpose of the analysis is to demonstrate whether a sample contains only one component an R of 1 or less can provide that information with a short analysis time. If the purpose is to quantitate from the chromatogram or to subject the column effluent to examination by mass spectrometry, then an R of 1.5 is preferable. If the intent is to collect or trap components from the effluent, then an R of 2 or greater, requiring longer analysis time, might be necessary.

Chromatographic parameters

Fig. 1.1. shows a number of measurements which can readily be obtained from a chromatogram. These can be used to help describe the processes operating in the chromatographic separation, and to elucidate those parameters which may be useful in controlling the separation. If the peaks are gaussian and W_{Ba} and W_{Bb} are equal, an alternate and equivalent to the above definition for resolution can be derived:

$$R = 1/4 \frac{k_b}{(k_b + 1)} \frac{(\alpha - 1)}{\alpha} (N^{1/2})$$

With this definition, resolution can be described as the product of a capacity term, a selectivity term, and an efficiency term.

The capacity term refers to the capacity of the sytem to retard the passage of a component through the column, not the gravimetric or volumetric capacity of the system. The capacity factor, or capacity ratio, k_b, is defined as the ratio of the time the component spends in the stationary phase, t', to the time it spends in the mobile phase, t_m:

$$k_b = \frac{t'_b}{t_m} = \frac{t_b - t_m}{t_m}$$

The capacity factor is related to the distribution ratio through a column geometry term, β, the phase ratio:

$$\beta = \frac{\text{Column volume of mobile phase}}{\text{Column volume of stationary phase}}, \text{ and}$$

$$K_D = \beta k$$

The phase ratio, β is a constant for a given column; therefore, those parameters which change K_D change k. At a constant carrier gas flow rate with a particular stationary phase K_D and k will only be affected by the column temperature. If column temperature is lowered there is more interaction of the component with the stationary phase, K_D and k increase, and retention time increases.

The second term in the general resolution equation is a selectivity term, and refers to the relative interaction of two components with the stationary phase. The selectivity, α, is defined as the ratio of the time component 'b' spends in the stationary phase to the time component 'a' spends in the stationary phase. Alpha is also related to K_D and k:

$$\alpha = \frac{t'_b}{t'_a} = \frac{K_{Db}}{K_{Da}} = \frac{k_b}{k_a}$$

Since the two components elute rather close together, their distribution coefficients are not very different; therefore the ratio of these distribution coefficients will be greater than, but close to, 1.0. Because of this, selectivity is quite insensitive to changes in column temperature. It is, however, sensitive to changes in type of stationary phase. The selectivity of a methylsilicone stationary phase for two components will generally be quite different than that of a polyethylene glycol stationary phase.

The last term in the general resolution equation is column efficiency, N, the number of theoretical plates. Column efficiency relates band broadening and time in the column for a particular component:

$$N = 16\left(\frac{t_a}{W_{Ba}}\right)^2 = 5.545\left(\frac{t_a}{W_{1/2a}}\right)^2$$

The term "theoretical plates" results because some of the early workers in GC were in the petroleum industry where distillation columns frequently have plates to promote liquid/vapor contact. A theoretical plate can be thought of as one liquid/vapor equilibrium step, comparable to a liquid/liquid equilibration in a separatory funnel where one of the phases may be separated after equilibration and then contacted with fresh second phase. Chromatography is a dynamic rather than equilibrium process; however, the theoretical plate concept is applicable in describing the mass transfer which occurs within the column.

Column efficiency is frequently expressed as H or HETP, the column height (length) equivalent to one theoretical plate, a reciprocal function of N:

$$H = \text{HETP} = \frac{L}{N} = \frac{\text{Column length}}{\text{Number of theoretical plates}}$$

The use of HETP has the advantage of expressing the column efficiency independent of total column length. In any particular column it is desirable to minimize HETP, thereby producing as many plates as possible for that column. It should be noted that N, or HEPT, is only meaningful as an indicator of efficiency if the chromatogram from which measurements are made is produced under isothermal conditions.

Van Deemter *et al.* (1956) described those factors which affect band broadening and therefore column efficiency:

$$H = A + \frac{B}{\mu} + C\bar{\mu} + D\bar{\mu}$$

where

$$\bar{\mu} = \frac{L}{t_{\mathrm{m}}} \text{ cm/s.}$$

In the van Deemter equation, $\bar{\mu}$ is the average linear carrier gas velocity, A is the eddy diffusion term, B is the molecular diffusion in the mobile phase term, and C and D, are the resistance to mass transfer terms in the stationary and mobile phases respectively. In order to minimize H it is necessary to optimize $\bar{\mu}$, and to minimize the A, B, C, and D terms.

The A, eddy diffusion or multipath, term contributes to band broadening because there are channels of various lengths through the packed column. This term can be minimized by using narrow diameter columns tightly and uniformly packed with smaller diameter packing. Typically a packed column will have an inside diameter of 2–4 mm, and be packed with 100/120 mesh packing. Once the column is packed the A term has a fixed contribution to efficiency; it is a constant independent of mobile phase flow rate. Since there are no packing particles in WCOT columns, the A term is zero.

The B, molecular diffusion in the mobile phase, term addresses the contribution to band broadening due to the tendency of component molecules to diffuse from higher to lower concentration areas in the mobile phase. It is minimized by tightly packing the column to decrease the time a component molecule spends between contacts with stationary phase; by using larger cross-section carrier gas molecules, i.e., nitrogen rather than helium; and by increasing carrier gas flow rate.

The *C*, resistance to mass transfer in the stationary phase, term deals with the mass transfer or diffusion of component molecules in the stationary phase. It may be minimized by using low stationary phase loading, e.g. 3%, to produce a very thin film of stationary phase and decrease 'pooling' where solid support particles contact each other; using a low viscosity stationary phase; operating at a temperature which will reduce the stationary phase viscosity; and using a slow carrier gas flow rate.

The *D*, resistance to mass transfer in the mobile phase, term is related to the rate at which component molecules can travel through the mobile phase between contacts with the stationary phase. It can be minimized by using smaller diameter columns; a lower cross-section carrier gas, helium or hydrogen rather than nitrogen; and by operating at a low mobile phase flow rate. In packed column GC this contribution is so much smaller than the other three terms that it can be neglected; however, it is extremely important in WCOT column GC.

There are obviously several decisions to be made in trying to minimize HETP. It would be desirable to use large carrier gas molecules to reduce the '*B*' contribution, but smaller ones to reduce the '*D*' contribution. Since the '*D*' term can be neglected in packed column GC, nitrogen is preferred as a carrier gas to obtain greater efficiency. In WCOT column GC hydrogen or helium are distinctly preferrable to nitrogen. In addition it would be desirable to operate with a high $\bar{\mu}$ to decrease the '*B*' effect, but a low $\bar{\mu}$ to decrease the '*C*' and '*D*' effects. The optimum flow rate could be determined by differentiating the van Deemter equation and setting it equal to zero. Practically it is done by calculating HETP under a variety of flow rates and

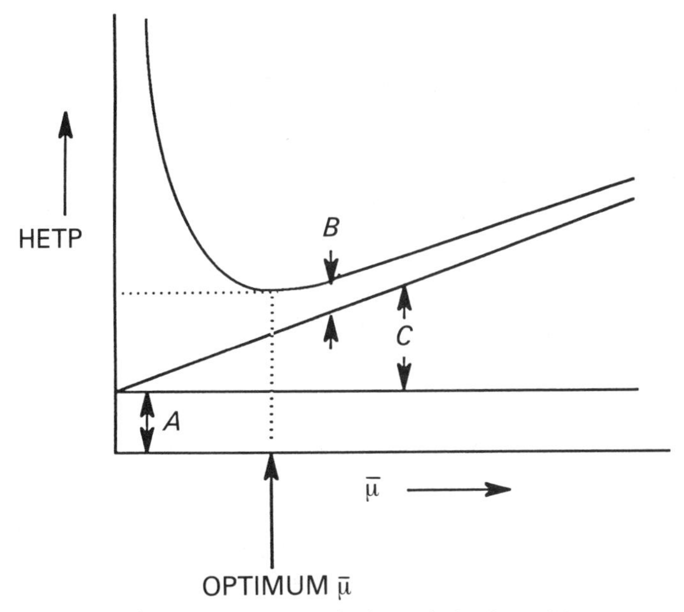

Fig. 1.3. van Deemter plot for packed column GC.

plotting a van Deemter curve as shown in Fig. 1.3.

At the minimum in the van Deemter curve HETP will be a minimum, therefore column efficiency is maximized. For a 2–4 mm inside diameter packed column this will be at a $\bar{\mu}$ of 8–10 cm. It is advisable to operate at a somewhat faster flow rate to decrease analysis time with little sacrifice in column efficiency. Operating at lower than the optimum can result in a rather large decrease in column efficiency.

Carrier gas is delivered to the injector end of the column at a constant pressure, typically about 20–25 psig for a packed column. The detector end of the column operates at atmospheric pressure. The pressure drop across the column, ΔP, is a direct function of carrier gas flow rate, F, carrier gas viscosity, U, and column length, L; and an inverse function of square of the cross-sectional area, A, available for gas flow:

$$\Delta P = \frac{(F)(U)(L)(Q)}{A^2},$$

where Q is a constant.

Unlike liquids, the viscosity of gases increases with increased temperature: therefore since ΔP, L, Q and A do not change, F must decrease as U increases with temperature. This requires that $\bar{\mu}$ be determined at the highest temperature anticipated for the program, so that it does not fall below the optimum as the program is run.

This briefly summarizes the parameters available to the analyst to evaluate and control his chromatography. In beginning an analysis the flow rate should be set to give a $\bar{\mu}$ 50–100% greater than the optimum at a temperature where it is suspected that the first peak of interest will elute. This may require a scouting run. The temperature can then be adjusted to give a capacity factor, k, of 2–10. Examining the basic resolution equation this means that the capacity term is contributing 67–91% of its maximum. A calculation of N will then afford a measure of the column efficiency. For a packed column this should be 1500–2000 plates/m; for a WCOT column N would be expected to be 3000 plates/m or greater.

With this start, the sample mixture may then be run to determine if resolution is acceptable. If the sample contains a mixture of early and very late eluting peaks, then a temperature program may be necesary to resolve the early peaks and elute the later ones with satisfactory shape, and in a reasonable time.

If the first peaks of interest have a k of 4–6, column efficiency is reasonable, and there are components which are not resolved, then the selectivity of the column may not be appropriate for the separation and a column with a different stationary phase may have to be considered.

WCOT column chromatography

A packed column 2 m long will usually require a head pressure of about 20–25 psig, and may have 4000 theoretical plates. If the column length is trebled to give about 12 000 plates the resolution may be increased by a factor of 1.73, and as can be seen from the pressure drop equation the operating pressure will be 60–75 psig. This is about the limit of packed column capability. WCOT columns consist of an open tube with stationary phase on the inner periphery. The pressure drop through a 15 m, 0.25 mm ID WCOT column will be about 4–6 psig, and there can be 45–60 000 theoretical plates available at a much lower operating pressure. Thus the WCOT column has much greater resolution capability. In addition the fused silica surface is much more inert than glass, resulting in less adsorption, therefore better peak shape.

The greater separating power of WCOT column permits the analysis of much more complex samples than is possible with packed columns. The lower operating pressure also allows the use of much longer columns, as much as 100 m.

To summarize the differences between packed and WCOT column chromatography:

(1) Column efficiency per unit length is much greater with WCOT columns, 3000–3500 compared to 1500–2000 plates/m.
(2) The higher column efficiency with WCOT columns results in more narrow peaks which is ideal for use with ionization detectors, but relatively poor for use with thermal conductivity detectors.
(3) The amount of stationary phase in a WCOT column is much less, therefore the amount of sample injected must be smaller.
(4) Pressure drop per unit length is much less for WCOT columns permitting the use of longer columns.
(5) Temperature response is quicker for WCOT columns because of their thin walls and low mass; therefore, oven temperature in programmed runs will more truly reflect the stationary phase temperature, and thermal equilibration is faster.
(6) Column stability and useful life are better for WCOT columns.

1.3 EQUIPMENT

Chromatographic system

To help in understanding the components of a GC system, and of particular help in trouble shooting a system, it is convenient to divide it into five functional areas:

— Carrier and support gas supply and control
— Injector
— Column and oven
— Detector
— Data system

Gas supply system

The purpose of the carrier gas supply system is to supply clean, dry, pure carrier gas to the injector under controlled, reproducible conditions.

For packed column GC the carrier gas of choice is nitrogen for use with FIDs, and helium for use with TC detectors. For WCOT column chromatography the carrier gas of choice is either helium or hydrogen. These gases as well as air and hydrogen for FIDs are usually supplied in approximately 200 cubic foot (STP) cylinders at an initial cylinder pressure of 2500 psig or slightly less. A two-stage pressure regulator is used to reduce the tank pressure to system pressure, 5–60 psig.

The carrier gas should be filtered through a series of traps to remove water, hydrocarbons and oxygen before introduction into the injector. Commercial traps are available from most chromatographic supply houses, and are usually useful for about three tanks of carrier gas before replacement or regeneration. The use of these traps is particularly important in WCOT column chromatography to minimize noise and interfering peaks in the chromatogram, and to prolong column life.

Gas is usually supplied to the injector at constant head pressure of about 25–30 psig for 2–3 m packed columns; and 4–6 psig for 10–15 m, 0.20–0.25 mm ID WCOT columns. A feedback flow controller is frequently used to maintain this pressure under varing flow conditions, and a pneumatic or electronic gauge is usually used to monitor the inlet pressure. If more than one GC is supplied from a gas manifold, intermediate pressure controllers are frequently required. It is important for the chromatographer to thoroughly understand the plumbing of his system in order to be able to control the carrier and other flows in a reasonable manner.

Injector

The purpose of the injector is to introduce a measured amount of sample from atmospheric into the system at operating pressure with as little band broadening as possible. Since chromatography always results in some band broadening it is essential that this be minimized in the injector. Sample measurement and introduction is most frequently done with a syringe and needle through a septum or other pressure seal, either manually or automatically for liquid or gas samples; or through a loop injector for gaseous samples.

A loop, or six-port gas sampling valve, permits introduction of the sample into a fixed volume sample loop at controlled pressure, and then introduction into the system at operating pressure. A schematic of the sample valve in the load position

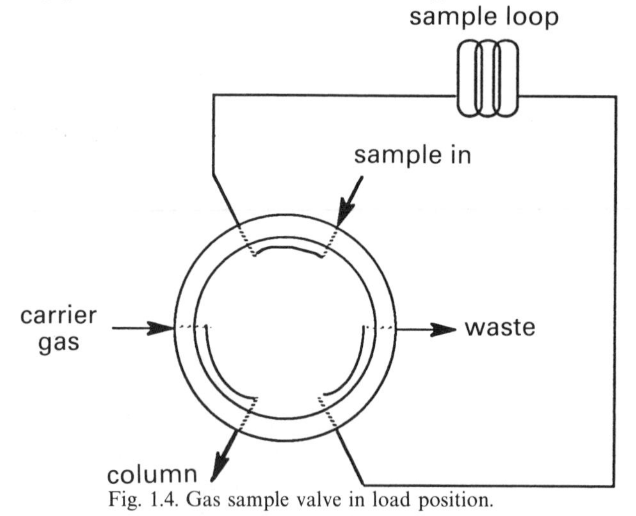

Fig. 1.4. Gas sample valve in load position.

is shown in Fig. 1.4. To inject, the inner portion of the valve is rotated 60 degrees clockwise so that the carrier gas port communicates with the sample loop and sweeps the sample onto column. Depending upon the sophistication of the system, and sample requirements, the valve loading and injection functions may be automatic or manual, and the valve may be equipped for temperature control and backflushing.

For gaseous samples this represents quite a reproducible injection system.

The injector most commonly used in WCOT column GC is the split/splitless

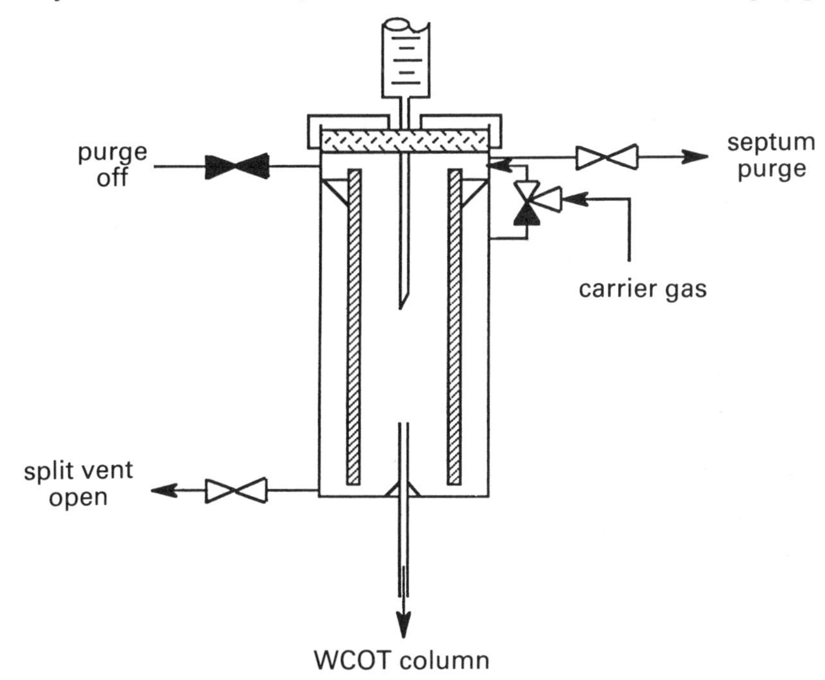

Fig. 1.5. Split-splitless injector in split mode.

injector as illustrated schematically in Fig. 1.5 in the split mode. The purpose of the septum purge is to sweep a low flow, 2–3 ml/min, of carrier gas across the septum and vent it to minimize the presence of extraneous peaks from septum bleed in the chromatogram.

In the split mode the purge valve is in the closed position and the carrier gas is directed to the top of the injector and down through the liner. Head pressure is adjusted to produce a flow rate of approximately 1 ml/min through the column; and the split vent valve is adjusted to give the desired split ratio, split vent flow:column flow. This permits the injection and vaporization of approximately 0.5–1.0 μl of sample, which can be conveniently measured, with only a fraction of the sample taken on column. For example, if 1.0 μl of sample is injected and vaporized, with a 99:1 split ratio, then a 0.01 μl aliquot of sample goes on column. With 0.20–0.25 mm ID columns it is desirable to have about 50–50 ng of the each component of interest on column. This results in a sufficient amount of detection without overloading the column.

With samples of widely varying volatilities it is recommended that split ratios be greater than about 25:1 to obtain a true aliquot on column, i.e., minimize discrimination against the lower volatility components. Injector liners frequently are designed to contain a small amount of packing, glasswool or low polarity stationary phase such as 10% methylsilicone, or a mechanical device to promote vaporization and mixing so that a representative split is obtained.

The primary advantages of the split mode of injection are that it is quite stable and reproducible; it is relatively free of solvent effects; and split ratio and column flow can be varied rather easily. If the split vent valve and column are thought of as two resistances in parallel, then it can be seen that, in the injector illustrated, the column flow can be adjusted, without affecting the split ratio, simply by adjusting head pressure. Alternatively, if head pressure is maintained constant, then the split ratio can be adjusted, by adjusting the split vent value, without changing column flow rate.

For samples of low concentration the injector may be operated in the splitless mode, which allows 80–95% of the sample injected to be directed on column. Figure

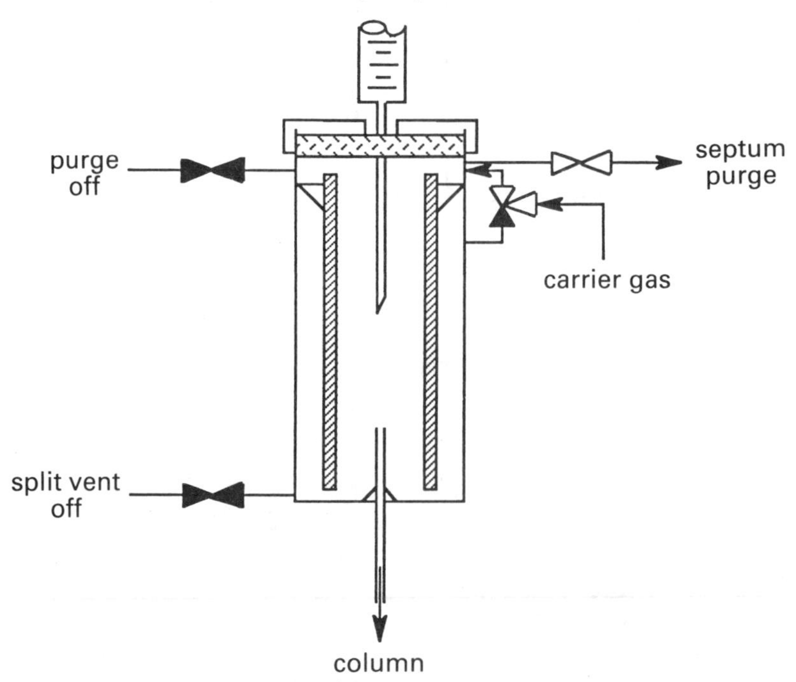

Fig. 1.6. Split-splitless injector in splitless: injection mode.

1.6 shows the injector in the injection phase. The split vent valve is closed, therefore all of the vaporized sample and solvent is conducted onto the column. By maintaining the column (oven) temperature about 20–30°C below the boiling point of the solvent, the sample components are focused at the head of the column by what has been described by Harris (1973) and Grob & Grob (1972, 1978) as the 'solvent effect'. After approximately 1.5–2 injector volumes of carrier have swept most of the sample onto column, usually 30–90 seconds, the valves are switched to the positions shown in Fig. 1.7, and the temperature program is initiated. At this time all sample in the injector (that which has not been directed onto column) is vented to prevent 'dribbling' the last 5–15% on to column, which would result in extraneous peaks or band broadening.

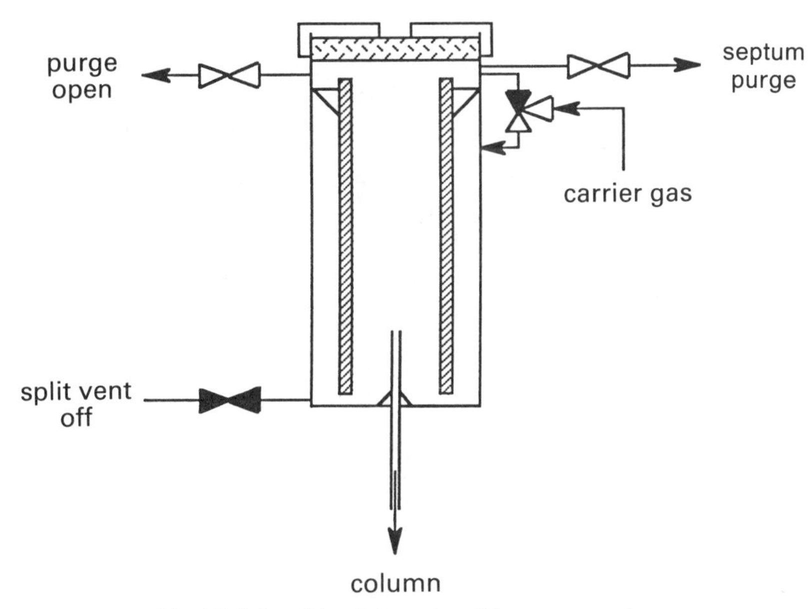

Fig. 1.7. Split-splitless injector in splitless: purge mode.

As the temperature program is initiated the solvent passes through the column rather rapidly, resulting in a large solvent peak; and the sample which has been focused at the head of the column begins to chromatograph.

The principal advantage of splitless injection is that it is applicable to samples of low concentration. The principal disadvantage is that solvent selection, temperature program, and time from injection to purge, are variables which must be optimized.

A third popular method of sample introduction in WCOT column GC is cold on-column injection. Fig. 1.8 shows a schematic of one such injector. In the loading phase the injector is cooled, in some designs with liquid nitrogen or carbon dioxide, the valve is opened, and the sample is deposited within the cold column. After deposition of the sample, the needle is withdrawn, the valve closed, coolant flow is stopped, the temperature of the injector is increased to vaporize the sample, and the temperature program is initiated.

The principal advantage of cold on-column injection is that the total sample is injected and there is no discrimination against lower volatility components as can occur with split injection. In addition the solvent effect does not have to be addressed as in splitless injection. The principal disadvantage is that automated on-column injection is somewhat more complicated.

Positioning of the column is critical with any injector. The manufacturer's recommendations should be followed initially, and then adjustments can be made based on experience. Once the best position is found, the distance from the column tip to the bottom of the retaining nut should be measured and recorded. When installing a column the nut and graphite ferrule can be positioned on the column and then several centimetres of column removed to remove any graphite shaved off

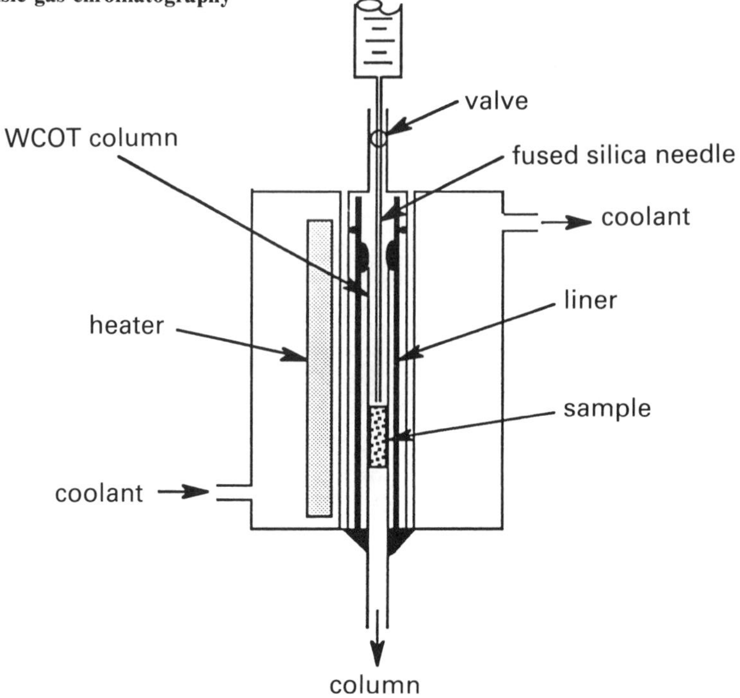

Fig. 1.8. Cold on-column injector.

the ferrule. The correct distance from the tip of the column can then be measured and marked with a small dot of typewriter correction fluid. This will facilitate reproducible installation of the column. The same procedure can be used in making the detector connection.

Column

At the center of the chromatographic system is the column, which separates the components of the sample. The most common types of columns are shown in Table 1.1. As indicated earlier, except for the separation of fixed gases or rather clean, simple separations such as forensic alcohol analyses, most separations are better addressed by WCOT column GC.

All of the WCOT columns except the super-bore, 0.75 mm ID, are fused silica; the super-bore columns are glass. Fused silica WCOT columns offer several advantages:

(1) The columns are much more inert than glass or metal columns; therefore peak shapes are generally better, i.e., less tailing and more nearly gaussian.
(2) Polymerized and bonded stationary phases are very stable and reproducible, resulting in low bleed and excellent column life.
(3) The columns' low mass allows quite fast and reproducible response to temperature changes. This, combined with stationary phase stability, improves the effectiveness of chromatographic characterization techniques such as retention indices.
(4) The fused silica columns are inherently straight, which facilitates installation.
(5) The columns are flexible and rather rugged; they can be coiled or manipulated

Table 1.1. Gas chromatographic columns

Designation	Internal dia. mm.	Typical flow rate ml/min
Packed:		
Analytical	2–4	30
Preparative	6–10+	60–100
WCOT		
Fused silica		
Micro-bore	0.1	0.25
Narrow bore	0.2–0.25	1
Wide bore	0.32	2–3
Mega-bore	0.53	6–10
Glass		
Super-bore	0.75	10–20

to conform to irregular geometries such as might be required to make connections to detectors such as a mass spectrometer or infra-red detector.

(6) The efficiency of micro, narrow, and wide bore WCOT columns allows the chromatographer to operate with fewer stationary phases.

(7) While the efficiency per unit length of mega-bore and super-bore columns may not be appreciably greater than packed columns, the longer lengths which can be used put many more theoretical plates at the analyst's disposal.

The principal disadvantages of WCOT columns is the limited amount of stationary phase present, which drastically limits the amount of sample which can be chromatographed. Narrow bore columns will most frequently have a stationary film thickness of 0.25 μm. To increase the sample handling capacity a film thickness of 1.0 μm is available. To further increase sample capacity wide bore columns with film thicknesses of up to 5 μm are available. It must be understood that increased film thickness increases retention time, and that increased column ID decreases column efficiency.

While it is possible to convert packed column instruments to narrow or wide bore operation, it is much more satisfactory to have an instrument specifically designed for operation with these columns. Alternatively it is rather inexpensive and convenient to convert packed column instruments to mega-bore or super-bore operation and take advantage of fused silica and/or longer length columns.

The choice of carrier gas is more critical with WCOT columns than with packed columns. As indicated previously, nitrogen will give better efficiency than helium or hydrogen in packed column GC; and this is also true in WCOT column GC. However, examination of Fig. 1.9 will show that while the efficiency, lower HETP, is slightly better at optimum $\bar{\mu}$ with nitrogen, the flow rate is low and the slope of the curve at higher flow rates is quite steep. With helium the optimum HETP occurs at about 20–25 cm/s compared to 8–10 cm/s for nitrogen. Thus by using helium rather than nitrogen it is possible to operate at 40–45 cm/s and obtain acceptable efficiency and faster analyses. With hydrogen as carrier the optimum occurs at about 40 cm/s, and it is possible to operate at 70–80 cm/s with quite good efficiency.

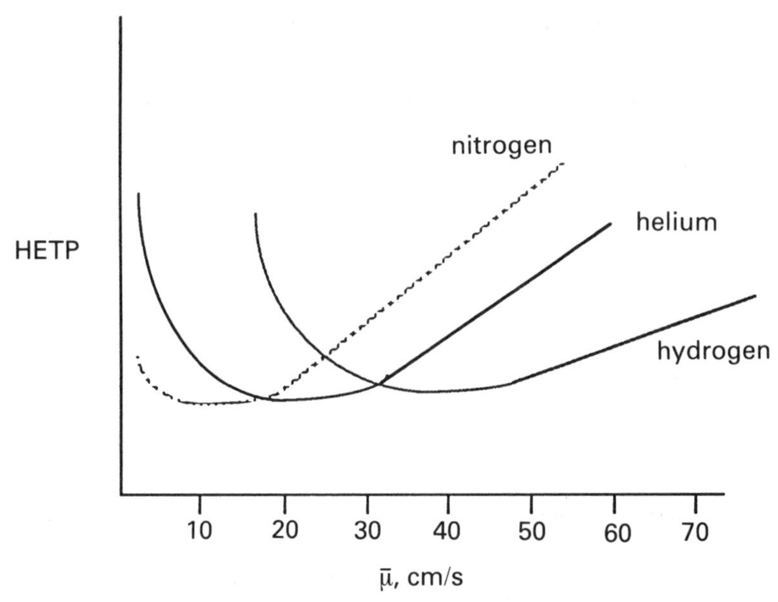

Fig. 1.9. van Deemter plots for WCOT column GC.

The majority of WCOT column GC analyses are done under temperature programming conditions. It is possible to set $\bar{\mu}$ at the higher values, 45, 80 cm/s for helium and hydrogen respectively, at the initial temperature and obtain better efficiency as the flow rate decreases with increasing temperature.

Column life can be expected to be in the range of weeks to years depending upon care. Longevity is promoted by use of effective water, hydrocarbon, and oxygen traps, proper installation and care, and good sample preparation techniques.

The flexibility and ruggedness of the columns facilitate installation; however, care must be taken to eliminate abrasion of the polyimide coating caused by rubbing against oven parts under the vibrations which are present in most oven designs. Proximity to any hot wires or other thermal stress can also result in degradation of the polyimide coating and column failure. If the system is not to be used for an extended time, over night, the temperature can be lowered to approximately 100°C, and with a 1 ml/min flow through the column and 2–3 ml/min septum purge there should be little or no accumulation of compounds which will decrease column life and result in 'trash' peaks when the system is next used.

Many samples must be recovered from complex matrices which can result in decreased column performance and life if introduced into the system. Generally, time and effort used in sample clean-up are well spent. Over time there will be materials which collect in the column, almost invariably at the head end. When chromatography deteriorates, with wide and/or distorted peaks and loss of efficiency, these accumulated compounds can be removed by removing the first 0.5 metre of column and reinstalling the clean part.

Care must be exercised in modifying column length. The column will break wherever the polyimide coating is disrupted. The polyimide coating can be scored at the desired spot by use of a sharp knife or a diamond tipped tool just as a piece of glass tubing is scored with a file where it is to be broken. Once the coating has been scored the part to be discarded should be held in a dependent position so that any small pieces of fused silica or polyimide do not fall back into the part of the column to be used. The end of the column should be examined to assure that the break is square and that there are no particles in the column. A 20 power magnifying glass can be effectively used for this purpose.

Detectors

The response of ionization detectors depends on the mass flow rate through the detector rather than being a function of the concentration of analyte through the detector as is the case with conductivity detectors. The efficiency of WCOT column GC results in sharp, narrow peaks, which means that the component has a high mass flow rate, ng/min, although a rather low concentration, ng/ml/min when going through the detector. For this reason flame ionization (FID), electron capture (ECD), and nitrogen/phosphorous (NPD) detectors are much more effective than a thermal conductivity (TC) detector.

With some detectors it may be necessary to use make-up gas to increase flow through the jet. If the detector is designed to operate with 30 ml/min of carrier gas and the column flow rate is approximately 1 m/min, significant band broadening can occur owing to dead volume, mixing, effects if make-up gas is not used. If it is required, nitrogen can be used at a flow rate of 30 ml/min, and it can be introduced with the hydrogen if there is not a specific port for it. Fig. 1.10 shows the basic components of an FID.

In the FID as the carrier (and make-up gas, if used) exits the column it is partly ionized in the flame at the jet tip. With a constant potential between the two electrodes a fixed current will flow in the circuit formed by the electrodes, the resistance gap between the electrodes, and the electrometer. The current due to carrier gas is balanced to give an electrometer output which is displayed as 'baseline'. As a sample component elutes from the column it is ionized in the flame, producing an increased number of ions in the gap. This reduced resistance in the gap results in an increased current since the potential is constant. The measurement and amplification of this current are output as the detector response and displayed as a deviation from baseline. As analyte mass flow rate increases, reaches a maximum, then decreases, a peak is formed.

The FID has a broad range of applicability, responding to most organic compounds which can be chromatographed. It does not respond, or responds very poorly, to compounds such as water, carbon monoxide, carbon dioxide, hydrogen sulfide, carbon disulfide, and fixed gases. It has good sensitivity and a linear dynamic range, and a ratio of limit of linear response to minimum detectable quantity, of about 10E6, which makes it very useful in quantitative analyses.

The use of specific detectors such as the NPD or ECD can significantly increase detectability and reduce the signal to noise ratio in the chromatogram. However, they are not a substitute for good sample preparation technique. Undetected

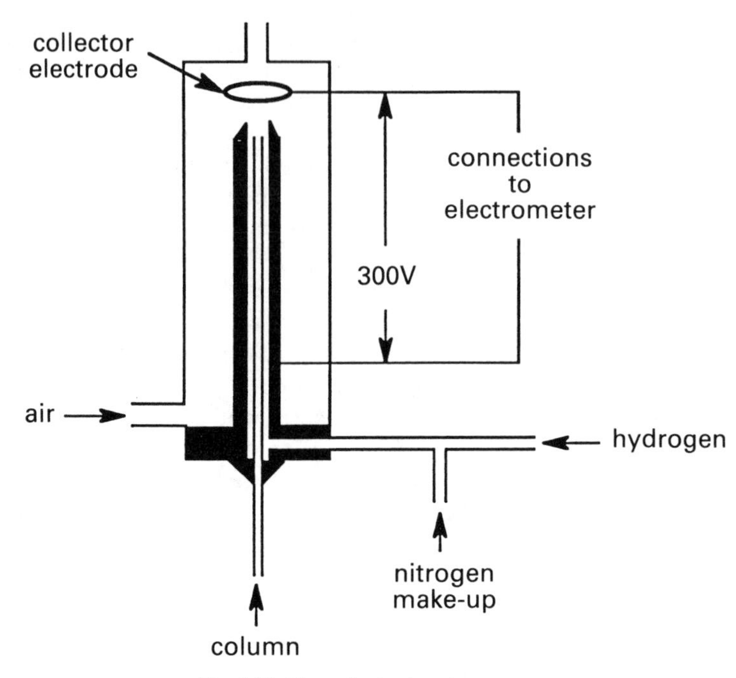

Fig. 1.10. Flame ionization detector.

components may interfere with the chromatography, and may be very significant if the sample is subsequently subjected to mass spectral or infra-red analysis.

In the ECD, radiation from a source such as nickel-63 ionizes part of the carrier gas, producing a standing current in the electrode gap. As components which have significant electrophilic substituents, e.g. halogens, enter the gap they 'capture' electrons, resulting in a decrease in the current in the electrode gap. This change in current is amplified and results in the presentation of a peak. The ECD has better sensitivity than the FID for electrophilic compounds, and very poor sensitivity for other components, permitting the selective detection of the electrophilic compounds.

The NPD employs a heated active element coated with an alkali metal, rubidium or cesium, above the flame at the jet tip. The alkali metal promotes the selective ionization of organic compounds containing nitrogen and phosphorus which are then detected. The flame is operated with decreased hydrogen and air flows, compared to an FID, to minimize ionization of non-nitrogen/phosphorus containing compounds.

A comparison of some commonly used GC detectors is shown in Fig. 1.11. Diagrams and operating characteristics of these and some other GC detectors are discussed by Hyver (1989) and Buffington & Wilson (1987).

The high resolution and the peak shape obtainable with WCOT column GC make it well suited to coupling with a mass spectrometer or Fourier transform infra-red detector to obtain molecular identification data. Fig. 1.12 is a schematic of a GC/MS system.

In GC/MS advantage is taken of the tremendous separating ability of GC and the

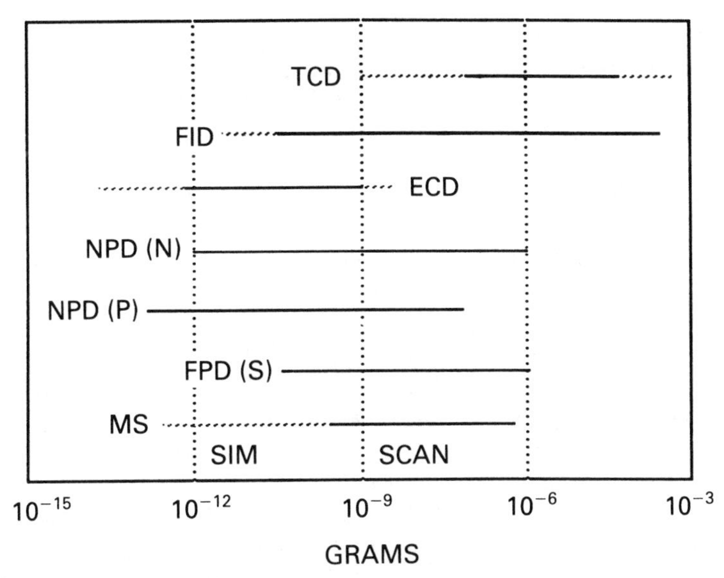

Fig. 1.11. Comparison of detector response.

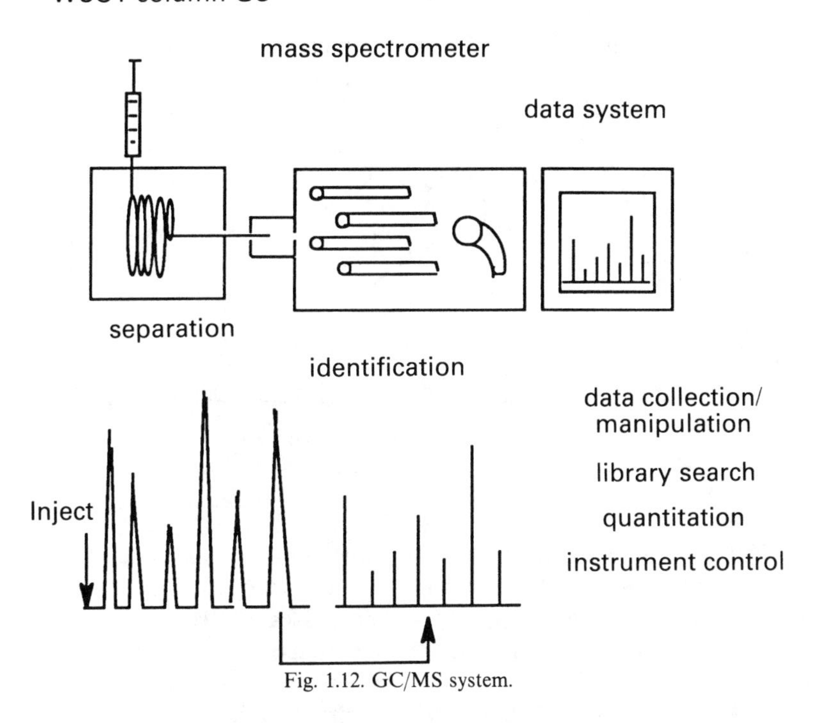

Fig. 1.12. GC/MS system.

molecular identification ability of MS. If a 0.20–0.25 mm ID WCOT column is used, the entire column effluent, about 1 mm/min, may be taken directly into the ion source of the mass spectrometer. If packed columns or larger bore columns are used, it is necessary to employ a separator between the GC and ion source in order to remove most of the carrier gas so as not to exceed the capacity of the MS vacuum system.

Since the mass spectrometer has no compound separating ability it is essential that only one molecular type be examined at any given time to eliminate mixed spectra which are generally nearly useless. Thus the very powerful analyical ability of the combined GC/MS system, especially when coupled with a computerized data system to handle the prodigious amount of information which may be obtained.

Ions are formed in the ion source by subjecting them to a high energy electron field, electron impact (EI) ionization, or ion molecule reactions with an ionized reagent gas (chemical ionization) (CI). The result in EI MS is ionization of the molecules of interest into positively and negatively charged ions and neutral particles. Depending upon the stability of these "molecular ions", further fragmentation may occur. In positive ion MS the conditions in the ion source are such that singly charged positive ions will be accelerated into the mass analyzer. This stream will consist of groups of fragments with differing mass, a single positive charge, and all with the same kinetic energy since they have been accelerated by the same potential.

This stream of ions may be separated into groups of like mass/charge by a mass filter. The mass filters most commonly used in forensic application are quadrupole mass filters, of which the ion trap is a special case; although some magnetic secter instruments are also used.

As each group of like mass ions exits the mass filter and enters the detector, a response proportional to the number of ions of each mass is generated and this information is captured by the data system. The result is often displayed in tabular

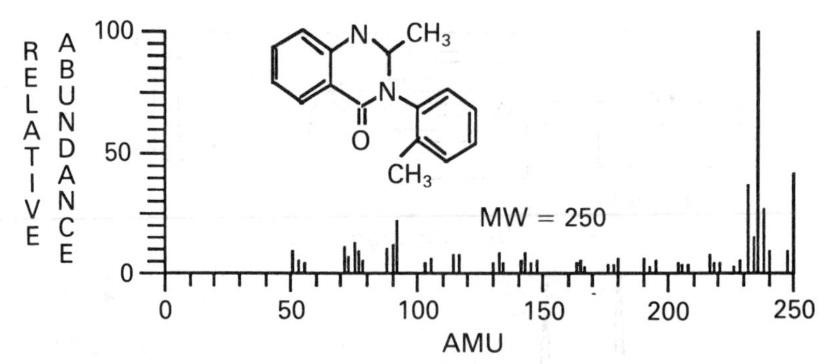

Fig. 1.13. Electron impact spectrum of methaqualone.

form or plotted as a frequency diagram, as shown in Fig. 13. This spectrum is plotted as a normalized frequency diagram with the most abundant ion, the "base peak", assigned a value of 100 and each of the other masses plotted as a percentage of the base peak. If the mass spectra are obtained under reproducible conditions — frequently 70 electron volts ionizing electron energy — with a properly tuned and calibrated MS system, then the spectra are reproducible. It is possible to compare with published and/or in-house generated reference spectra for identification and to use a library search routine to find the spectra which most closely match the sample spectra. These can then be examined more closely to confirm identification.

In CI MS a reagent gas, typically isobutane or methane, is present in much higher concentration than the sample component; therefore, the reagent gas is preferentially ionized. Through ion-molecule reactions with the sample component molecules several charged species may be obtained. Since the ion-molecule reactions are much less energetic than EI ionization there is much less fragmentation. There may be the transfer of a proton to the component molecule which results in a mass one dalton greater than that of the molecule, the M + 1, or pseudomolecular, ion. There may be an adduct formed between the reagent gas ions and the component molecule which results in ions of discrete numbers of daltons greater than the molecular weight. The overall result is a much simpler spectrum, with much less information than obtained from EI MS; however, the significant feature is that the M + 1 ion is increased in intensity and the higher ions are accentuated, giving an indication of the molecular weight. The lower mass fragment information is not present.

Examples of EI and CI MS are shown in Fig. 1.14. A discussion of the design and operating characteristics of MS components and systems is presented by McFadden (1973), and examples of the application of GC/MS are given by Watson (1976). Mclafferty (1973) presents a very good introduction to mass spectral interpretation.

Recently, GC-FTIR detectors have become commercially available. These have the ability to provide identification information where the mass spectra of two or more compounds may not be sufficiently different to unequivacally identify them. There has been some very effective use in analysis of isomers and other structurally related compounds; however, their popularity has been slow to develop.

Data system

WCOT column GC results in very fast eluting and in narrow peaks, frequently only seconds wide. In order to capture the large amount of data which is quickly generated, it is necessary to employ a microprocessor or computer based data system. These can vary from a relatively simple recording integrator to a full scale, multi-instrument laboratory data system. The choice of data system largely depends upon the laboratory's needs and fiscal situation. Another consideration is the amount of time and technical ability available to effectively use larger systems. It is essential that this time and talent be assessed in selecting a system. Because of the vast range of systems available the laboratory can find the system(s) which match its requirements; however, the education and acquisition process may be quite lengthy.

Chromatographic characterization

Several systems have been devised to characterize chromatographic separations. The simplest method is to use absolute retention time. For most situations this is not satisfactory since retention time will vary with operating temperature and flow rate, column length and amount of stationary phase, and reproducibility of injection.

An improvement is to use relative retention time, the ratio of the retention time(s) of the compound(s) of interest to a marker compound present in the same chromatogram. While this is better than absolute retention time, it still does not facilitate comparison of data from different laboratories.

Beginning with the Clausius–Clapeyron equation (Barrow 1966) it can be demon-

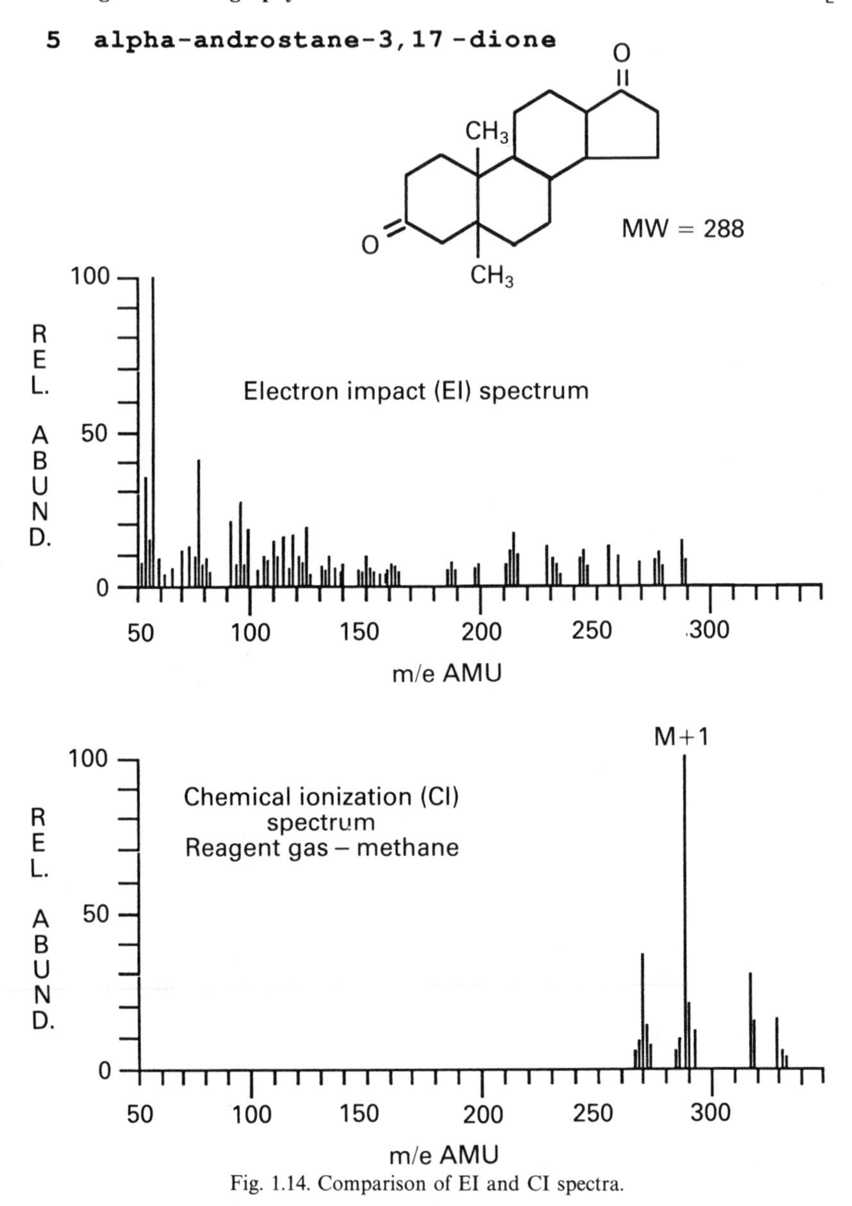

Fig. 1.14. Comparison of EI and CI spectra.

strated that for a homologous series of organic compounds chromatographed under isothermal conditions the logarithium of the adjusted retention time is a linear function of carbon number. For a different homologous series a linear function with a different slope will be obtained. As a result of his doctoral work, Kovats (1958) published the concept of retention index (RI). He chose a series of normal aliphatic hydrocarbons as his markers since they were readily available, stable, and chromatographed nicely, and a large number of them were available. He defined RI as the carbon number multiplied by 100, e.g., the RI for *n*-hexane is 600. By chromatographing these markers he obtained the log t'_r vs RI function, either graphically or as an equation. A compound of interest can then be chromatographed under the same conditions, and, using its t'_r and the function obtained from the *n*-alkane markers, a RI could be determined. If compound X has a RI of 982 it chromatographs on that particular stationary phase as if it were a *n*-alkane with 9.82 carbon atoms.

The advantage of the Kovats RI is that it is dependent only on stationary phase type, and is independent of column length, mobile phase flow rate, and stationary phase film thickness or loading. Within a 10–20°C temperature range it is also independent of column temperature. This means that a data file or library of RIs can be compiled and used to make a preliminary identification of compounds in the chromatogram. It also allows meaningful inter-laboratory exchange of data.

Kovats RIs are obtained under isothermal conditions, and most WCOT column GC is done under temperature programmed conditions. Van Den Dool & Kratz (1963) published studies of RIs obtained under temperature programmed conditions. Under temperature programmed conditions the retention time is a nearly linear function of RI, and is dependent on the stationary phase and temperature program. Halang *et al.* (1978) published a cubic spline data treatment which described the t_r vs RI function quite well. A much simpler treatment was published by Anderson (1981). In this method the RI is determined by linear extrapolation between two bracketing hydrocarbon markers as shown in Fig. 1.15. This has been shown to be equally as effective as the cubic spline treatment if a sufficient number of markers are used.

Several investigators (Perrigo & Peel 1981, Marozzi *et al.* 1982, Anderson & Stafford 1983) have demonstrated the use of the RI technique in toxicologic and drug analyses. The use of RIs allows the chromatographer to compile his own library of RIs or to use published data if he uses the same stationary phase and temperature program as the published data. Stafford (1985) has shown that for dimethyl polysiloxane WCOT columns the RIs are independent of column manufacturer and whether they are coated or crosslinked.

The use of retention indices improves the chromatographer's ability to make a presumptive identification from the chromatographic information, and guides him in establishing conditions for confirmation. It also promotes inter-laboratory use of information.

A concept proposed as an alternative to theoretical plates for assessing column efficiency is the Trennzahl, or separation, number, *TZ*, described by Kaiser (1963). *TZ* is equal to the number of peaks of similar size and shape which will fit between two consecutive markers with a resolution of 1.177. This represents a separation of 4.7 standard deviations of the peak width, and represents greater than 98% separation. It is calculated as shown in Fig. 1.15. The use of *TZ* allows the chromatographer to readily monitor efficiency, and also allows him to predict whether a column will separate two compounds if he knows their RIs. For example, if a column has a $TZ = 9$, then there are ten equal intervals between two adjacent *n*-alkane markers. Since there is a difference of 100 RI units between these markers, then two compounds which have a 10 unit difference in RI can be nearly baseline separated on that column.

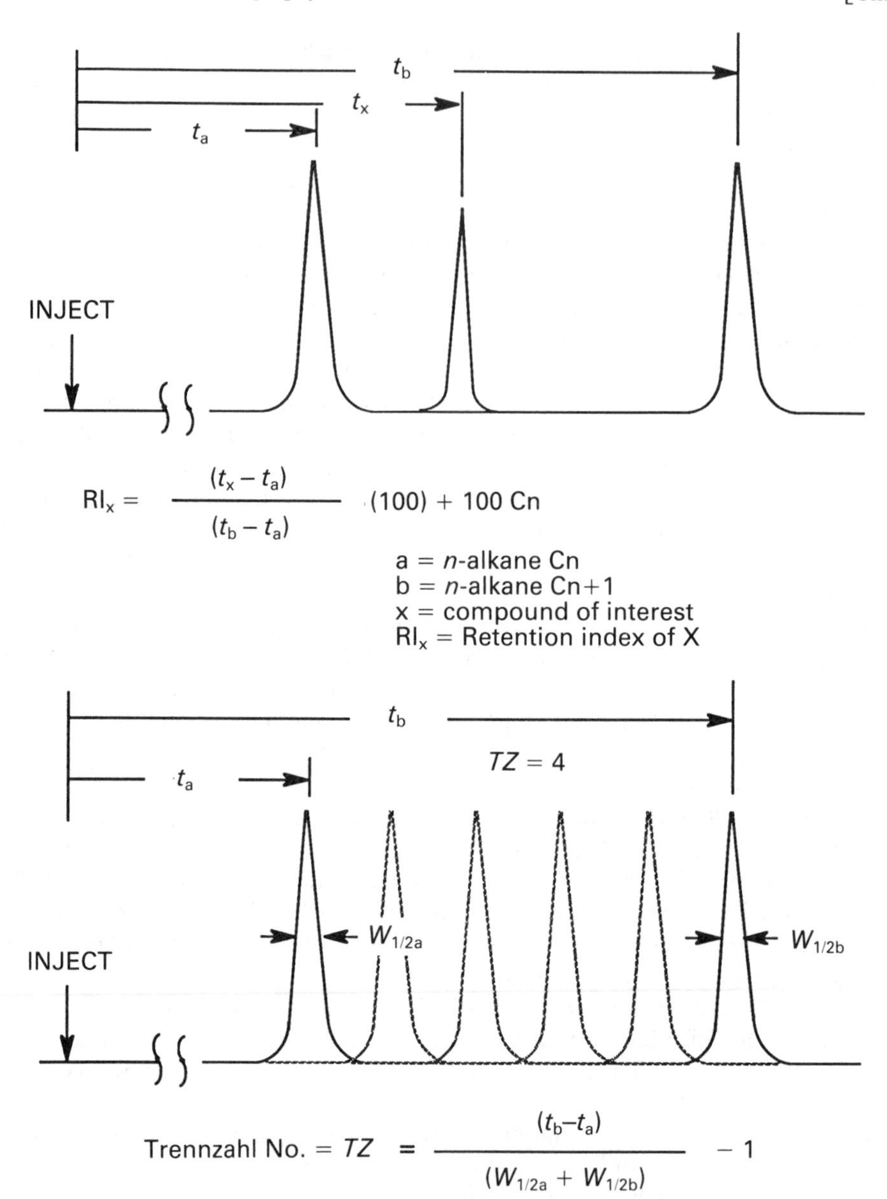

Fig. 1.15. Retention index and trennzahl number.

Conversly, if the difference in RI is known, it is possible to calculate the *TZ* necessary to separate the two compounds.

TZ differs from *N* in that it can be determined under actual temperature program conditions and can give a more meaningful estimate of column performance. On a dimethyl polysiloxane column many drugs elute in the *n*-alkane C-15 to C-20 region. A 15 m, 0.25 mm ID WCOT column should be capable of a *TZ* of 16–20 in that region.

Both retention index and Trennzahl number are valuable aids in characterizing and monitoring separations and column performance.

Applications

The powerful separating ability of GC, particularly WCOT column GC, and its ability to interface with molecular identification instrumentation, make it ideally suited to many forensic analyses from accelerant analysis to toxicology. It would be redundant to include specific examples here when any chromatography supply house catalog will contain many more numerous illustrations; and of course any forensic journal dealing with the analysis of organic compounds is likely to have examples more nearly matching the reader's interests.

Gas chromatography is the most widely used analytical technique in forensic laboratories, and with increasingly more useful instrumentation and methodology its use will continue to grow. It should be understood that, as with any powerful tool, the user must be well versed in the fundamentals of the technique in order to control and use it to the greatest advantage.

REFERENCES

Anderson, W.H. (1981), Drug detection in biological specimens, Doctoral thesis, University of Tennessee.

Anderson, W.H. & Stafford, D.T. (1983), Application of capillary gas chromatography in routine toxicological analyses, *J. High Resolut. Chrom. and Chrom. Commun.*, **6** 247–54.

Barrow, G.W. (1966), *Physical chemistry, 2nd ed.*, McGraw-Hill, New York.

Buffington, R. & Wilson, M. (1987), *Detectors for gas chromatography – a practical primer*, Hewlett-Packard Co., Avondale Division, Box 900, Avondale PA, 19311.

Dandeneau, R.D. & Zerenner, E.H. (1979), An investigation of glasses for capillary chromatography, *J. High Resolut. Chrom. and Chrom. Commun.*, **2** 351–56.

Dandeneau, R.D. & Zerenner, E.H. (1990), The invention of the fused-silica column: an industrial perspective, *LC/GC*, **8** 908–912.

van Deemter, J.J., Zuiderweg, F.J. & Klinkengerg, A. (1956), Longitudinal diffusion and resistance to mass transfer as causes of nonideality in chromatography, *Chem. Engr. Sci.*, **5** 271–89.

Gluechauf, E. (1955), *Ion exchange and its applications*, Soc. of Chem. Ind., London, 34.

Golay, M.J.E. (1957), Theory and practice of gas liquid partition chromatography with coated capillaries, in *Gas chromatography*, Coates, V.J. *et al.*, eds, Academic Press, New York, 1–13.

Golay, M.J.E. (1958), Theory of chromatography in open and coated tubular columns with round and rectangular cross-sections, in *Gas chromatography*, Desty, D.H., ed., Butterworths, London, 36–55.

Grob, K. & Grob, G. (1972), Methodik der Kapillar-Gas–Chromatographie Hinweise zur vollen Ausnutzung hochwertiger Saulen: Teil 1, Die Direkteinspritzung, *Chromatographia*, **5** 3–12.

Grob, K. & Grob, K., jr. (1978), Splitless injection and the solvent effect, *J. High Resolut. Chrom. and Chrom. Commun.*, **2** 57–64.

Halang, W.A., Langlais, R. & Kugler, E. (1978), Cubic spline interpolation for calculation of retention indices in temperature-programmed gas-liquid chromatography, *Anal. Chem.*, **50** No. 13 1829–32.

Harris, W.E. (1973), Some aspects of injection of large samples in gas chromatography, *L. Chrom. Sci.*, **11** 184–187.

Hyver, K.J. (1989) *High resolution gas chromatography, 3rd. ed.*, Hewlett-Packard Co., Avondale Division, Box 900, Avondale, PA, 19311.

James, A.T. & Martin, A.J.P. (1951), Liquid-gas partition chromatography, **48** vii.

James, A.T. & Martin, A.J.P. (1952), Gas-liquid partition chromatography, *Analyt*, **77** 915–32.

Kaiser, R.Z. (1963), *Anal. Chem.*, **189**.

Keulemans, A.I.M. (1957), *Gas chromatography*, Reinhold, New York.

Kovats, E. (1958), Gas-chromatographische Charakterisierung organischer Verbindungen: Teil 1. Retentionsindices aliphatischer Halogenide, Alkahole, Aldehyde und Ketone, *Helv. Chem. Acta*, **41** 1915–32.

McFadden, W.H. (1973), *Techniques of combined gas chromatography/mass spectrometry: applications in organic analyses*, John Wiley, New York.

McLafferty, F.W. (1973), *Interpretation of mass spectra, 2nd. ed.*, W.A. Benjamin, Reading, MA.

McWilliam, R.G. & Dewar, P.A. (1958), in *Gas chromatography*, Desty,. D.H., ed., Butterworths, London, 142.

Marozzi, E. *et al.* (1982), Use of retention index in gas chromatographic studies of drugs, *J. Anal. Tox.*, **6** 185–92.

Martin, A.J.P. & Synge, R.L.M. (1941), A new form of chromatography employing two liquid phases, *Biochem. J.*, **35** 1358–68.

Perrigo, B.J. & Peel, H.W. (1981), The use of retention indices and temperature programmed gas chromatography in analytical toxicology, *J. Chrom. Sci.*, **19** 219–26.

Pretorius, V. (1979), in *75 years of chromatography*, Ettre, L.S. & Zlatkis, A., eds, Elsevier, Amsterdam, 335.

Ramsey, W. (1905), A determination of the amount of neon and helium in atmospheric air, *Proc. R. Soc. London*, **A76** 111–14.

Stafford, D.T. (1985), Comparison of retention indices on bonded and coated dimethyl polysiloxane capillary columns, *Crime Lab. Dig.*, **12** No. 3 60–63.

Tswett, M. (1906), Physikalisch-chemische Studien uber das Chlorophyll. Die Adsorption, *Ber. Dtsch. Bot. Ges.*, **24** 316–23; English Translation, Adsorption analysis and chromatographic methods, *J. Chem. Educ.*, **44** 4 (1967) 238–42.

Van Den Dool, H. & Kratz, P.D. (1963), A generalization of the retention index system including linear temperature programmed gas-liquid partition chromatography, *J. Chrom.*, **11** 463–71.

Watson, J.T. (1976) *Introduction to mass spectrometry: biomedical, environmental and forensic applications*, Raven Press, New York.

2

Gas chromatography and forensic drug analysis

A. Karl Larsen, B.A.
Illinois State Police, Suburban Chicago Laboratory, Maywood, IL, USA

Donna Wielbo, Ph.D.
Department of Pediatrics, University of Illinois at Chicago, Chicago, IL, USA

2.1 INTRODUCTION

The ever increasing problem of drug abuse in the United States and North America during the 1980s and the beginning of the present decade (Hindmarsh & Opheim 1990) is well recognized and has been a major concern of recent government administrations (Montagne 1990). Drug abuse trends are influenced by availability and accessibility of the substance. The ease of obtaining and ingesting a drug is influential in deciding the nature and duration of a drug trend (Montagne 1990). Owing to the escalating drug problem, the analysis of drugs of abuse now accounts for a major proportion of the workload of major city crime laboratories, being in excess of 70% of cases handled (Jackson 1986). The following drugs are the most commonly encountered: cannabis, cocaine, phencyclidine, heroin, and amphetamines, as well as the amphetamine derivative designer drugs (Meyers pers. com., Larsen pers. com., Katzung 1989c, Gold & Giannini 1987, Hoffman 1983a-d, Giannini 1989, Miller & Gold 1989, Gold 1989). These compounds may be encountered in their pure form or in street preparations adulterated with various compounds such as caffeine, strychnine, and sugars (Jackson 1986, Hoffman 1983d, Shannon 1988, Schawbell 1990, Barnfield *et al.* 1988) which may complicate the analysis. In the United States, cocaine street samples presently have a range of purity from 14% to 75% with an average of about 40% (Shannon 1988). Heroin samples are usually less than 2% pure, whereas phencyclidine and amphetamine are commonly encountered in their pure form (Meyers pers. com., Larsen pers. com.). Unfortunately, automated methods used in clinical drug identification and therapeutic drug monitoring are not directly applicable to the identification of illicit dosage forms submitted to the crime laboratory. Forensic examination of street drugs, that is the analysis and identification of controlled substances to the elucidation of questions that occur in judicial proceedings, is time consuming and considerably complicated, requiting the use of several tests.

Tests currently employed generally include an initial screening of the sample to detect or exclude the presence of controlled substances. This usually entails the use of nonspecific but highly sensitive color tests. Having sensitivities in the low microgram range, spot tests as such give an indication of a class of compounds but not specific compounds within that class (Jackson 1986, Clarke 1955, 1961, 1962, Fulton 1969, Stevens 1986). Color tests have the advantage of extreme simplicity and require neither elaborate equipment nor special skills. The most widely used color tests are those employing concentrated sulfuric acid, either alone or as a solvent for some other compound, usually an oxidizing agent. The test is carried out by adding a drop of the reagent to a minute portion of the solid alkaloid and noting the subsequent color change (Clarke 1962). The next screening method often entails the use of thin layer chromatography (TLC) (Jackson 1986). This technique can separate a mixture of compounds of the same class of drugs, or can differentiate between unknown but different classes of drugs. Preliminary tests may also be performed using ultraviolet (UV) spectrophotometry (Jackson 1986), which again is a nonspecific technique but which gives information concerning the general structure of the compound. If preliminary tests prove to be positive, then the presence of specific drugs must be confirmed, and as a legal requirement in most states, the structure of the illicit substance must be proven.

Elucidation of the compound's structure is usually achieved by mass spectrometry (MS) coupled to a gas chromatograph, or by infra-red analysis (IR) (Jackson 1986, Raversby & Gorski 1989, Chapman 1986, Raverby 1987). Since street drugs are rarely encountered in their pure form, it is usually necessary to extract the drug of interest from any interfering compounds or diluents before GCMS or IR analysis. The extractions, either liquid-liquid or solid phase (Raverby 1987, Smith & Stewart 1981), may be complicated, entailing extraction, filtration, and evaporation stages (Jackson 1986), which increase the anlaysis time for a single sample. With these methods of analysis the time taken to analyze one sample can take from 90 minutes to 4 hours.

2.1.1 Drugs of abuse

2.1.1.1 Cannabis

Cannabis has been widely available for many years. This drug, which has sedative hypnotic properties, is generally found in one of three forms: marijuana, hashish, or hash oil. Marijuana is plant material from the plant *Cannabis sativa* which contains the active principles cannabinol (CBN), cannabidiol (CBD) and tetrahydrocannabinol (THC) (Fig. 2.1) (Turner 1981). The fruiting and flowering tops of the plant are particularly rich in these compounds. The plant is usually dried, mixed with tobacco

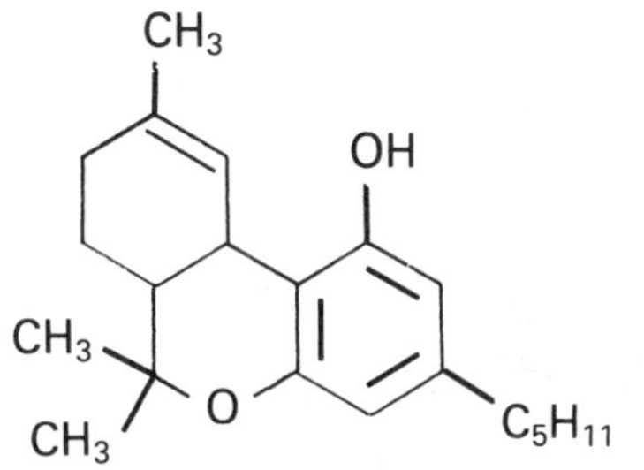

Fig. 2.1. Structural formula of tetrahydrocannabiol (THC).

and smoked. A resinous material, high in cannabinoid content, can be extracted from the tops of the cannabis plant. This is dried and pressed into blocks and is known as hashish. If the plant material or hashish is extracted with gasoline, or other volatile solvent, which is then evaporated to low volume, a thick oily liquid is produced. This liquid, hash oil, consists of around 70% THC and is the most potent form of the drug. Hash oil is usually smoked by placing a few drops onto a regular cigarette (Jones *et al.* 1976). THC is rapidly metabolized in the body, and the principal urinary metabolite is 11-nor-delta-9-THC acid (THCA). THCA can be detected in the urine 4–6 days after cannabis use, and up to 30 days in chronic users.

The physiological effects of marijuana use include increased pulse rate, hypotension, drowsiness, visual disturbances, impaired balance, loss of concentration and poor coordination. Cannabis usage has been reported to be associated with at least 20% of motor vehicle accidents in the United States (Giannini *et al.* 1985). The long term consequences of cannabis use have been the subject of a great deal of research, with conflicting data and subsequent heated debate. However, the drug has been reported to be associated with genetic (Linn *et al.* 1983), immune, reproductive (Kolodny *et al.* 1974), pulmonary, cardiovascular, gasteroenterologic, and neurologic disorders. As with other psychoactive drugs, chronic cannabis use has been reported to result in paranoia, psychosis and personality disturbances. Of the latter there appears to be a direct link between cannabis and the so called amotivation syndrome (Klah *et al.* 1989).

2.1.1.2 Cocaine

Cocaine is a naturally occurring alkaloid from the plant *Erythroxylon coca*, grown in South American countries in the Andes regions (Gold & Giannini 1987, Hoffman 1983a, Loper 1989, Johanson & Fischman 1989). Coca leaf chewing is a daily occurrence amongst the Andes indians (Jones *et al.* 1976), where the stimulant effects of the plant decreases fatigue during physical work.

Cocaine, a benzoic acid ester of ecgonine (Fig. 2.2), is a nervous system stimulant and local anesthetic drug (Katzung 1989c). Its abuse potential is enforced by the euphoric effects experienced by abusers a few minutes after administration. Cocaine's low cost, availability and ease of administration (Gold & Giannini 1987, Hoffman 1983a Johanson & Fischman 1989, Van Dyke & Byck 1982), presently labels it the most abused drug in the United States. Approximately 20 million Americans have

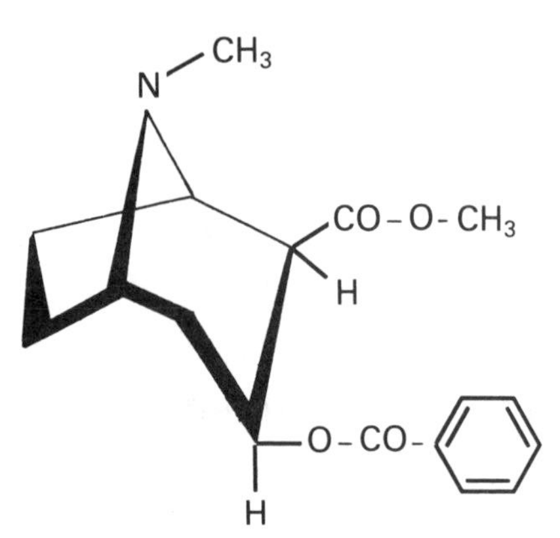

Fig. 2.2. Structural formula of cocaine.

tried cocaine at least once; in the region of 5 million abuse it regularly and 4 million try it for the first time each year (Bagasra & Forman 1989). In 1979 the National Narcotics Intelligence Consumers Committee (NNICC) estimated that between 25 000 and 31 000 kilograms of cocaine entered the United States. This increased to between 40 000 and 48 000 in 1980 and 178 000 kilos were imported in 1987, based on consumption data (Montagne 1990, Van Dyke & Byck 1982). It has been estimated that the amount of coca now being cultivated is 8–10 times greater than the amount needed to fulfill current demand for both licit and illicit cocaine (Montagne 1990). Previously considered a nonaddictive drug, this misconception has now been realized. Cocaine abuse is physiologically addictive, but the symptoms manifest as psychological disturbances (Loper 1989, Johanson & Fischman 1989, Van Dyke & Byck 1982, Fischman 1988). The most common route of administration is intranasal (IN) by insufflation or snorting; however, the recent appearance of free base cocaine, known as ‘crack’ or ‘rock’, made by alkalinizing the salt and extracting into nonpolar solvents, now allows the drug to be smoked. This route of administration produces a quicker onset of effects and more intense euphoria, but largely, the effects experienced are the same (Bagasra & Forman 1989): garrulousness, excitability, increased endurance to work, loss of inhibitions, talkativeness followed by depression and a feeling of persecution and paranoia with outbursts of aggression (Gold & Giannini 1987, Hoffman 1983a, Fischman 1988, Siegel 1982, Galvin 1988, Katzung 1989a). Other less commonly used routes of administration, less commonly encountered, for cocaine hydrochloride is by absorption across the rectal and vaginal mucosal membranes, by the gastrointestinal tract after ingestion and by intravenous (IV) injection after dissolution in water.

Death has been reported after IN, IV, intravaginal and free base use, and is largely dose dependent. Most complications arise from the cardiovascular effects of cocaine and its effects on blood pressure and cerebral circulation (Fischman 1988, Wetli & Wright 1979). The increase in recreational use in pregnant females has resulted in an increase in spontaneous abortions, *abruptio placentae*, and the occurrence of cerebrovascular accidents in premature newborns of cocaine-abusing mothers, as well as decreases in birth weight and head circumference (Dixon & Bejar 1989, Chasnoff *et al.* 1986).

The route of administration determines the bioavailability of the drug; 100% of IV cocaine is absorbed and 20–40% of an ingested or IN dose (Gold & Giannini 1987, Hoffman 1983a, Resnick *et al.* 1978, Javaid *et al.* 1978, Van Dyke *et al.* 1978, 1976). Cocaine undergoes a large first pass hepatic metabolism which accounts for the low bioavailability of ingested and insufflated doses (Jatlow 1988). Vasoconstriction of the blood vessel in the nasal mucosa also limits the absorption of the drug by the IN route (Jatlow 1988). Regardless of the route of administration of cocaine, repeated usage results in the development of tolerance to some of the effects of the drug (Ambre *et al.* 1988).

In the body, cocaine is rapidly metabolized to a number of inactive compounds. These include benzoylecgonine produced by a de-esterification reaction as a result of spontaneous hydrolysis of cocaine at physiological temperature and pH, and ecgonine methyl ester a product of de-esterification by plasma and liver cholinesterases subsequently metabolized to ecgonine by nonenzymatic hydrolysis. Benzoylecgonine is also further metabolized to ecgonine by plasma cholinesterases. The only active metabolite, Norcocaine, is produced by N-demethylation by liver enzymes. These metabolites are used as markers of cocaine usage, having longer biological half lives than cocaine (45–90 minutes) (Moffat 1986); they can be found in body fluids such as urine and blood, the body tissues such as liver and brain, days after cocaine usage. In toxicological samples the major metabolite, benzoylecgonine, is generally encountered in urine in μg/ml concentrations as long as 48 hours after use. Very little unchanged cocaine is excreted in urine. Blood concentrations are dose dependent. (Van Dyke & Byck 1982, Jatlow 1988, Moffat 1986a, Fulton 1962).

Cocaine is encountered on the street cut with a number of diluents and additives. These include the local anesthetics procaine and lidocaine which mimic the numbing sensation of cocaine when the taste test is employed. Sodium bicarbonate, starch, talcum powder and sugar are added to the sample to add bulk (Hoffman 1983d, Shannon 1988, Schawbell 1990, Johanson & Tischman 1989, Van Dyke & Byck 1982). Crack is produced in the form of a yellow, waxy lump, hence the street name "Rock".

2.1.1.3 Amphetamines

Amphetamine, a potent sympathomimetic amine, has very similar effects to cocaine, owing to its similar mechanism of action, the potentiation of catecholamine neurotransmitter effects, and its related anorectic, stimulant and cardiovascular effects (Hoffman 1983a, Miller & Gold 1989, Katzung 1989a). The simple chemical structure of the drug allows derivatives to be easily synthesized. These structural analogs have pharmacological effects like amphetamine, resulting in an increase in abuse and addiction potential. Essential to the pharmacological action of amphetamine is the unsubstituted aromatic ring, the two carbon side chain and alpha methyl and amine groups (Fig. 2.3A). The effect of amphetamine on the dopaminergic systems of the brain is well documented, and amphetamine psychosis is probably better understood than that of cocaine psychosis (Hoffman 1983a, Miller & Gold 1989). It is the effects

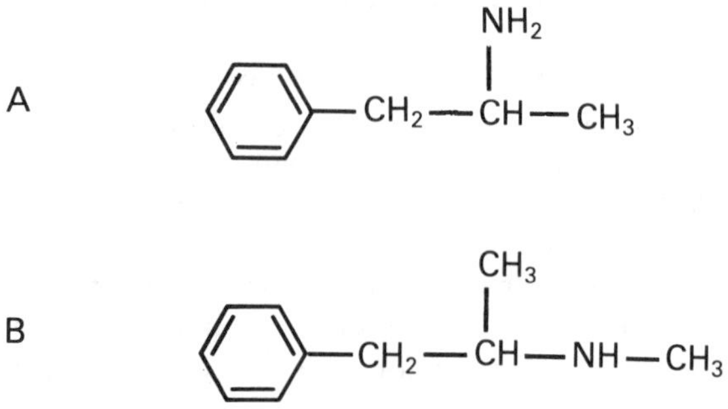

A $Ph{-}CH_2{-}CH(NH_2){-}CH_3$

B $Ph{-}CH_2{-}CH(CH_3){-}NH{-}CH_3$

Fig. 2.3. Structural formula of (A) Amphetamine, (B) Methamphetamine.

of amphetamine on the pleasure centers of the brain which confer its addictiveness. The euphoric effects and mood elevation are considered by addicts to be worth the secondary effects of depression, anxiety and insomnia as well as the psychiatric disorders often encountered by the abuser. The hedonistic effects of the amphetamines drive abusers to the addiction stage rapidly. A pattern of use has developed in which the drug is thought to enhance sexual pleasure, with sexual intercourse reported as markedly prolonged and orgasm enhanced. However, with chronic use sexual urges tend to disappear completely, a reaction also experienced with cocaine (Katzung 1989a). With long term abuse of amphetamine, induction of psychosis has been observed, particularly in persons with a history of psychosis. The pscyhotic person may become paranoid, socially withdrawn and functionally incapacitated. He also develops delusions of persecution and, because of this, readily becomes violent and dangerous. This type of psychosis is closely related to that of schizophrenia.

Amphetamines were first manufactured as racemic mixtures used initially as nasal decongestants. Today they are used medically in the treatment of hyperactivity in children, better known as attention deficit syndrome, as appetite suppressants and in the treatment of narcolepsy.

Tests of pharmacological activity of the two isomers of amphetamine have shown that D-amophetamine has greater activity than L-amphetamine. The pharmacological effects of D-amphetamine and the racemic mixture have been shown to be identical.

Methamphetamine, a methyl derivative of amphetamine, (Fig. 2.3B), is more potent than amphetamine as a central nervous system stimulant (Hoffman 1983a, Miller & Gold 1989). Oral abuse of the drug tends to lead to intravenous use with increasing addiction. The appearance of "ice" on the market, the free base methamphetamine, is now turning the route of administration to inhalation by smoking. Amphetamines are rapidly absorbed orally and rectally, upon direct contact with mucosal membranes, and from injection sites. Amphetamines undergo five routes of metabolism: aromatic hydroxylation, aliphatic hydroxylation, oxidative deamination, N-oxidation, and nitrogen conjugation. About 50% of a dose is metabolized by deamination in the liver and the remainder is excreted in the urine unchanged. Excretion by the renal system can be enhanced by acidifying the urine and slowed by alkaline conditions (Fulton 1962). Amphetamine has a plasma half life of 4–12 hours if the urine is acidic and 12 hours in patients where the urine is uncontrolled (Moffat 1986b). Methylamphetamine is readily absorbed after oral administration, 70% of which is excreted in the urine within 24 hours. Up to 40% is excreted as unchanged drug and 5% as amphetamine, the major active metabolite. The plasma half life of methylamphetamine is 9 hours (Moffat 1986c).

2.1.1.4 Phencyclidine

Phencyclidine, a synthetic phencyclohexylamine derivative (Fig. 2.4), initially developed as a general anesthetic, is now used therapeutically as a veterinarian preparation.

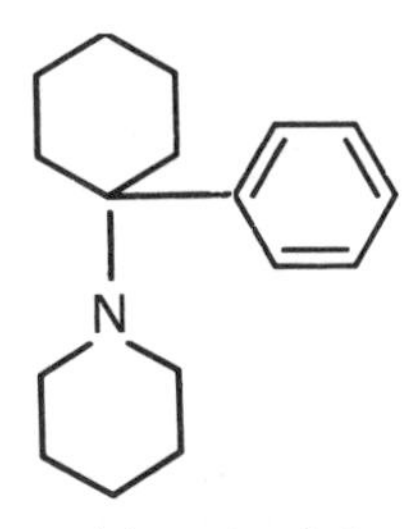

Fig. 2.4. Structural formula of phencyclidine (PCP).

Illicit phencylidine was first seen in table form but is now sold as a powder which is either snorted or smoked (Gold & Giannini 1987, Hoffman 1983b). The effects of phencylidine are experienced as slurred speech, nystagmus, a feeling of intoxication, a rolling gait and numbness of the hands and feet. Muscular rigidity, sweating, amotivation and staring may also be experienced. With higher doses more pronounced analgesia occurs, and hostile and aggressive behavior develops. Phencyclidine possesses both cholinergic and anticholinergic properties and has direct mycoardial depressant abilities (Gold & Giannini 1987, Hoffman 1983b, Schawbell 1990, Katzung 1989a).

PCP is absorbed easily when administered by all common routes of administration, including the digestive tract. In the acidic environment of the stomach and in acidic urine the drug with a pKa of 8.5 is in its ionized form, which prevents it crossing lipophilic barriers. This results in the accumulation of PCP in the stomach and urine. In the intestines the pH of the environment is such that the drug is largely un-ionized and highly lipophilic in nature. This results in enterohepatic circulation, which may account for the prolonged effects of the drug. Like most drugs, response is dependent upon the route of administration. Regardless of route the biological half life is variable and effects may last from a few hours to a week. The plasma half life of phencyclidine is between 7 and 46 hours (Moffat 1986d). The drug undergoes hydroxylation and glucuronidation during metabolism and is largely excreted by the renal system. About 50% of the dose appears as unchanged PCP in the urine. The rate of excretion can be enhanced by acidifying the urine (Gold & Giannini 1987, Hoffman 1983b).

2.1.1.5 Opiates

Opiates are derived from the latex extracted from the opium poppy, *Papaver somniferum* originating in Asia minor. Only two commercially available opiate drugs are naturally occurring, morphine and codeine. Heroin is the most widely abused semi-synthetic opiate, first synthesized, in 1874, by the relatively simple process of acetylation of the two hydroxyl groups present on morphine, (Fig. 2.5).

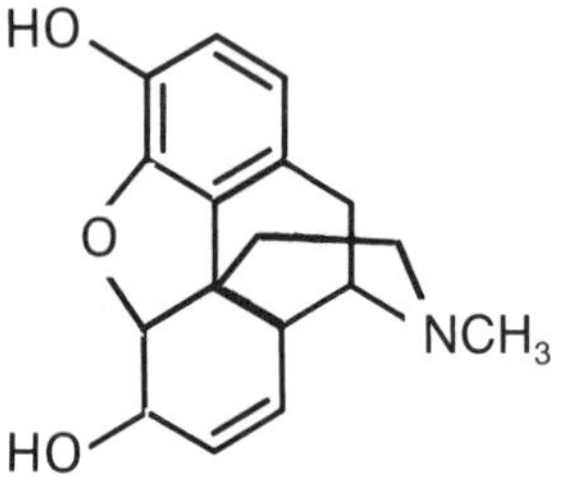

HO HO O NCH_3 A

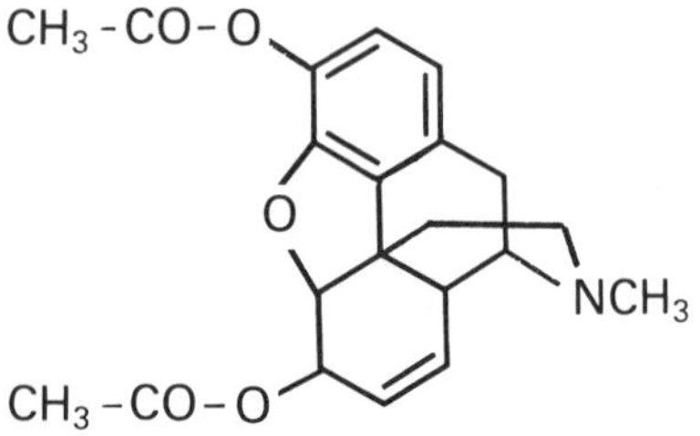

CH_3-CO-O CH_3-CO-O O NCH_3 B

Fig. 2.5. Structural formula of (A) morphine, (B) heroin (diacetylmorphine).

Owing to the lipophilic nature of heroin the pleasurable effects produced have been stated to be far superior to those of morphine which is less lipophilic in comparison. The lipophilicity of the drug determines the ease with which it crosses the blood-brain barrier and exerts its central nervous system (CNS) effects. Pharmacologic differences have been said to exist between heroin and morphine. Heroin has a greater analgesic potency than morphine, the effects of heroin administered by IV injection are perceived more rapidly than those of morphine administered by the same route, and the duration of effects of heroin do not last as long as those of morphine (Hoffman 1983c, Gold 1989, Katzung 1989a,b).

Heroin is not well absorbed after ingestion and is usually injected intravenously (IV), although there are reports of it being snorted in the same way as cocaine. Chronic inhalation by this route generally results in nasal septum perforation. (1983c).

Heroin is usually seen as a bitter tasting white crystalline powder and is mixed with diluents in illicit street samples. Diluents and adulterants may include sugar, caffeine, strychnine, quinine, and even brick dust and cleaning agents such as Ajax (Hoffman 1983d, Shannon 1988). In such street samples heroin may compose only 1% by weight of the total sample (Meyers pers. com., Larsen pers. com.). Many of the adulterants are cellular poisons, such as strychnine, which is readily absorbed from both the gastrointestinal tract and nasal mucosa. Strychnine is also a competitive antagonist of the CNS neurotransmitter glycine. Quinine has always been a common heroin adulterant with toxicity directed toward the heart, CNS and kidneys (Shannon 1988).

The effects of the drug are dependent upon a number of factors including: the amount of drug taken, the route of administration, the time span between doses and the degree of tolerance which the abuser has acquired through use.

After IV administration of heroin the following effects are commonly experienced: an immediate abdominal thrill, a decrease in blood pressure and some bodily itching, euphoria, sedation and mental clouding. Anxiety and worry are absent along with the sexual desire and hunger. First time users may actually experience dysphoria and vomiting, but with regular use the feeling of euphoria is soon experienced. IV administered heroin has a short biological half life of 3 minutes and is rapidly metabolized to 6-monoacetyl morphine, an active metabolite, reportedly equipotent with heroin. The second metabolic step is the hydrolysis of the second cetyl group, which occurs much more slowly and produces the end product of morphine, also pharmacologically active. The morphine formed, which has a longer half life of 2–3 hours, is then metabolized to form morphine-3 and morphine-6-glucuronide, excreted as a glucuronide conjugates (Hoffman 1983c, Gold 1989, Moffat 1986c). Other reactions include N-demethylation, O-methylation and N-oxide formation.

Codeine, structurally similar to heroin and morphine, is therapeutically used as an analgesic and cough suppressant, usually administered orally. The pharmacological effect of codeine is largely the same as that of morphine but is very much less intense. The euphoric effects of codeine are mild and it is therefore not widely abused. Abusers can be categorized into two groups, namely, experimenting teenagers and therapeutic addicts who have become addicted with continual use of codeine as a therapeutic agent. However, it is thought that heroin addicts, having difficulty obtaining their next fix, will abuse codeine as a means to alleviate withdrawal symptoms (Hoffman 1983c, Gold 1989). Codeine is well absorbed after oral administration, with a bioavailability of 50%. It is metabolized in the liver to form morphine by O-demethylation and norcodeine by N-demethylation. It is then conjugated to form glucuronides and sulfates. The plasma half life is 2–4 hours (Moffat 1986f).

Besides risk of fatal overdose, a number of other serious complications are associated with opiate dependence. Hepatitis B and human immunodeficiency virus (HIV) are complications of sharing hypodermic syringe needles. Bacterial infections lead to septic complications such as meningitis, osteomyelitis and abscesses of various organs. Homicide, suicide and accidents are more prevalent among heroin abusers than in the general population (Katzung 1989a).

The analysis of heroin seizures has always been a difficult matter. As it reaches the consumer the heroin is usually adulterated and diluted. Small percentages of heroin must often be separated from very complex mixtures (Moffat 1986a, Lerner & Mills 1963). Heroin at its point of origin is far from pure, as it usually contains some acetylcodeine as well as impurities of unconverted morphine and partially converted monoacetylmorphine owing to imperfect acetylation (Moffat 19862a).

2.1.1.6 Drug adulterants and diluents

Drug trafficking procedures result in marked adulteration and dilution of drugs from synthesis or cultivation to actual street sales. Of abused drugs, cocaine and heroin are the most extensively adultereated compounds on the street (Shannon 1988). Diluents and adulterants are chemical compounds which are purposely added to illicit street drug powders to dilute the active ingredients and increase bulk; these types of adulterants are usually inert compounds (Hoffman 1983d, Shannon 1988). Some active adulterants may be present as the result of contamination from the clandestine manufacturing process, or may be added because of similar pharmacologic and physiochemical properties to the substance of abuse. Five categories of adulterants are well recognized: sugars, stimulants, toxins, local anesthetics and inert compounds (Shannon 1988).

Identification of adulterants and chemical profiling of illicit drug samples allows interpretation of matching samples to be used in local cases of drug dealing and also provides intelligence information on the origin and network of samples which may be of use to provide major trafficking conspiracies (Barnfield *et al.* 1988). The greater the number of parameters used to compare the samples, the more significance the results derive.

Sugars are the most common diluents; life cocaine they are white, odorless crystals with minimal taste. Sugars are also cheap and easy to obtain. Inert compounds also used include inositol, a water soluble vitamin which is an isomer of glucose, and is converted to glucose in its metabolism. Inositol is essential in the synthesis of phospholipids and has little clinical toxicity other than being very irritating to the nasal mucosa.

Talcum powder or magnesium silicate, an inorganic mineral, is also used as an adulterant. The substance is white and crystalline in appearance, odorless, tasteless, cheap, readily available and without significant pharmacological effects. Starch, another common inert adulterant, is also odorless with an indistinct taste and is also metabolized to glucose *in vivo* (Hoffman 1983d, Shannon 1988).

2.2 GAS CHROMATOGRAPHY

The nature of mixtures of illicit drugs, diluents and adulterants submitted to the forensic laboratory generally necessitates the extraction or separation of the controlled substance from any compounds which may potentially interfere with the subsequent analysis. Gas chromatography has always been an integral part of the forensic drug chemist's arsenal, and nowadays is primarily employed as a means of separating the components of a drug mixture before they pass into a mass spectrometer for absolute identification. Improvements in equipment, columns, packings and supports over the last twenty years have enabled us to perform analyses which go far beyond the mere preliminary identification of drugs and their quantitation. We can now use the gas chromatograph as a tool to help in the identification of reaction pathways used to clandestinely manufacture drugs of abuse, characterize samples and help determine possible origins, and in some cases, determine the stereochemistry of the compounds in question. The relatively new technique of GC-FTIR-MS (gas chromatography-Fourier transform infra-red spectrophotometry-mass spectroscopy) makes possible the full characterization of an unknown compound, based on structural data as well as retention time information.

Forensic drug chemists have come to rely heavily upon gas chromatography for routine quantative and qualitative analyses required in the performance of their duties. This section will review the use of gas chromatography in the examination of drugs of abuse and is dedicated to those who have made the life of the drug analyst easier.

The main components which make up a gas chromatograph have been described in Chapter 1. Briefly, separation of injected compounds is achieved by using a column which is kept in a temperature controlled oven and through which a carrier gas, usually nitrogen, passes. The columns range from glass packed columns to different types of capillary columns, with several different packings or stationary phases. Once separation has been achieved on one of these columns the compounds pass into a detector. The available detectors include nitrogen-phosphorous (NPD), electron capture (ECD), flame ionization (FID), and the mass spectrometer (MS). The most popular of these for forensic drug analysis are the flame ionization detector (FID) and the mass spectrometer (MS). The FID, while not as sensitive as the ECD or NPD, is suitable for drug examination since sample size and therefore sensitivity is generally not a major consideration. Its ease of operation and maintenance make it the ideal workhorse for the volume of analyses which must be performed. The mass spectrometer, which is very sensitive though more complex, gives the analyst structural information about the compound under examination. This, together with retention data obtained from the gas chromatograph, can be used for the conclusive identification of unknown compounds.

In the development of assays for the examination of street drugs, the main consideration is in the choice of column. This section will therefore begin with a discussion of the types of column available and follow with information on the gas chromatographic analysis of the various classes of compound which forensic drug chemists most often encounter.

2.2 Further reading

This section is simply an overview of gas chromatography and its application to forensic drug analysis. For specific and more detailed descriptions of particular analyses, the following three manuals are recommended: Clarke's *Isolation and Identification of Drugs*, *Instrumental Data for Drug Analysis*, and the AOAC *Official Methods of Analysis*. Drugs are treated individually in these works, and if presumptive identification has been obtained, they offer valuable help in the determination of the nature of a compound. Additional current information can always be obtained from journals such as *Microgram*, *Journal of Forensic Sciences*, *Journal of the Forensic Science Society*, and *Journal of Chromatography*.

2.3 COLUMNS

2.3.1 Glass columns

Even as little as ten years ago, the glass packed column was widely used for gas chromatography. This column, varying between 2 mm and 4 mm in internal diameter and in length from 1 to 4 m, was filled with a support coated with stationary phase. The efficiency of these columns was by today's standards very low, being in the range of 2000 theoretical plates. Nonetheless, while the efficiency was low, the resulting analysis which they could give was, and still can be, excellent.

Packed columns have to be well maintained and cared for, but are relatively low in cost. These columns allow injection volumes of 10 μl or more and typically use a carrier gas flow rate of 30 ml/min or more. They give the analyst the ability to modify the stationary phase by simply repacking the column with bulk packing. In some cases, maintenance of the column can be extremely simple. When deterioration in column efficiency is noted, the column can be removed and may require only the replacement of the first few inches of the packing material. The available packing materials provide a wide range of analysis capability while having to keep only a few glass columns on hand. Further, an analyst can vary the analytical capabilities of the column by packing the column with one type of packing material while reserving a few inches at the injector end for a different type of packing which could help in the distinction of difficult to separate compounds. Cleaning procedures allow these columns to be used repeatedly or until broken.

Although packed columns are still used today for certain analyses, they have largely been replaced in the drug chemistry laboratory by capillary columns. Micropacked columns were a step between packed and capillary columns. These were longer (up to 25 m with a smaller internal diameter, and were up to 25 times as efficient having close to 50 000 theoretical plates.

2.3.2 Capillary columns

Capillary columns have in recent years been receiving much interest and are used in much of the ongoing research in chromatographic separation today. The first capillary columns were made of glass which made them extremely delicate and subject to fracture problems which can be frustrating to the analyst. Many times the sudden lack of sample eluting from the column could be traced to a break hidden somewhere in the 25 m or more column in the oven. Even more troublesome were the small holes which were worn in the side of a column by vibration. This results in a marked decrease in sample reaching the detector. The capillary columns of today are composed of much more durable fused silica. The stationary phase in these columns is either coated directly onto the internal wall of the column, the so called wall coated open tubular (WCOT) column, or is coated onto a support which is then held to the inside of the column, yielding the support coated open tubular (SCOT) column. The former of these is the more popular of the two and has rapidly increased in popularity and general use throughout the chromatographic community. Though these columns are expensive when compared to a glass packed column, they can have efficiencies exceeding 100 000 theoretical plates.

Today, most gas chromatographs come equipped for at least one capillary column. Those which do not can be adapted at a later time. Adaptations are also available for the conversion of some older instruments. The injection volume in these columns at 0.1 to 0.3 μl is much smaller than that of packed columns, and carrier gas glow rates are reduced to about 2 ml/min. As previously stated, the newer fused silica columns are sturdy enough for day to day use, and crosslinking of the stationary phase helps prevent column bleed and prolongs the life of the column.

2.4 SUPPORTS

The most used supports are variations of diatomaceous earths. Ideally, the support which the stationary phase is coated on should be nonreactive, have a high surface area to the volume ratio, be sturdy enough that it will not crush when packed into a column, and be stable under GC conditions. The diatomaceous earths fit these requirements. Composed of silica, this material is in reality the remains of microscopic plants called diatoms. The silica skeletons are subjected to high temperatures to increase their ability to resist breakage. If this heating is accompanied with a sodium bicarbonate flux, the resulting white support is called, among other names, Chromasorb W on Celite. If no such flux is used, the pinkish material which is formed is among other things called Chromasorb P or Firebrick (Jack 1984). Further treatments such as acid washing are used to improve the ability of the support to remain inert and be coated with stationary phase. Sizing the particles to make the support as uniform as possible completes the preparation for later coating of this material with a stationary phase. While other supports exist, they are not often used in forensic drug analysis. These can include various polymer supports and glass beads which have been treated.

2.5 CHOICE OF STATIONARY PHASES

Many different stationary phases have been used throughout the history of forensic drug analysis, though a large number have only limited use for specific analyses. The idea that one stationary phase can be used for all drugs of abuse was once thought to be improbable. The compounds which the forensic drug analyst sees, run the full range of polar to nonpolar and acidic amphoteric and basic. If there is one factor which runs in the scientist's favor, it would be that when compared to toxicological cases, there is a high ratio of drug in comparison to the matrix. One problem with the examination of street drugs, however, is that the matrix is composed of materials, some of which are closely related to the drug of interest, an example being cocaine which may be cut with other local anesthetic agents such as procaine or lidocaine. The stationary phases chosen for analysis must be versatile enough to accommodate a large diversity of drug types while being able to distinguish closely related drugs. The use of Kovats' work with retention indices, Trennzahl's concept of separation numbers and McReynold's constants, helps the forensic scientist to characterize the stationary phases and helps him to choose those which will enable the most efficient analysis to be done in the laboratory.

2.5.1 Kovats' retention indices

The original work by Kovats' (1961) has been described in detail in Chapter 1. Simply, the retention index (RI) is determined by using elution times of a series of *n*-paraffins and calculating the RI of a particular compound based upon the two hydrocarbons which bracket it.

$$I_X = 100 \left[Z + \frac{\log V_{NX} - \log V_{NZ}}{\log V_{N(Z+1)} - \log V_{NZ}} \right]$$

where Ix = retention index of compound X
Z = Number of carbon atoms in the first n-alkane standard
V_{NX} = retention volume of X

V_{NZ} = retention volume of first *n*-alkane standard
$V_{N(Z+1)}$ = retention volume of second *n*-alkane standard

Many studies have been done with this procedure and scientists who have compared data generated in different laboratories have found that interlaboratory results can be correlated (Stowell & Wilson 1987, Clark 1978). The formula can be used in either isothermal or temperature programmed analysis, with packed, or various types of capillary columns (Stowell & Wilson 1987, Japp *et al.* 1987a,b) with consistently linear relationships between the hydrocarbon length and retention times, and calculated RI values for drugs which are very close in these analyses. The best measure of RI is always that performed in-house, using the equipment and conditions indigenous to the particular laboratory.

The fact that different analyses using various sets of conditions, equipment, and analysts can generate similar numbers could allow the scientist to make a preliminary determination of an unknown, based on literature values before confirmation on the in-house equipment. Tables of calculated RI values for drugs can be found in many journal articles published over the last 20 years. These are too numerous to cite individually; however the work by Ardrey & Moffat (1981) lists compiled RI values for 1318 drugs on both SE-30 and OV-1 columns and is worthy of particular mention.

2.5.2 McReynolds constants

The McReynolds constants stem from work begun by Rohrschneider (1966). These two scientists used the RI values of compounds to characterize the performance of different stationary phases in gas chromatographic columns, the idea being that one could compare the RI value obtained on one column of a particular stationary phase with that observed on a different phase. This can be used to predict whether two different phases will react to compounds in a similar way, or if they are very different. The original papers by McReynolds (1970) or Rohrschneider (1966) can provide those with specific interest in this subject all of the information they desire.

2.5.3 Separation number

The separation number can be used to determine the number of fully resolved peaks that can fit between two neighbors in a homologous-*n*-paraffin series. In most drug analyses, this number will usually be low, enabling the scientist to be confident that the observed peak is the compound being sought (1986).

2.5.4 Stationary phases

Use of the above constants can help one to choose the correct column for a special analytical problem if the scientist wishes to start from scratch. Fortunately, many scientists have previously researched the problems encountered with most drugs and have described their findings. While drug analysis cannot as yet be accomplished with one perfect column, the research completed to date does suggest dimethyl polysiloxane (SE-30, OV-1, DB-1, HP-1 and other trade names) as the favored nonpolar stationary phase. Methylphenyl polysiloxane (SE-54, OV-17 and other trade names) in differing ratios of phenyl to methyl groups forms the most popular phases in both low and intermediate polarity phases. When the ratio is about 5% phenyl to 95% methyl, the polarity remains low and the stationary phase is sold under the names HP-5, DB-5, SE-52 amongst others. A ratio of 50% phenyl to 20% methyl results in an intermediate polarity phase and is generally referred to as HP-17, DB-17. High polarity phases are composed of polyethylene glycol derivatives and Carbowax derivatives known as Carbowax, DB-WAX and others (Both 1990).

The popularity of columns and phases depends to a large extent on the versatility and durability of the unit. The polysiloxanes serve well because of their very wide operational temperature range (−60 to 300°C). They are also inert and the columns are highly efficient. The Carbowax-type phases are sensitive to degradation by air and water when operated at moderately high temperatures (above 160°C) (Both 1990). This problem has resulted in the practice of derivatizing polar compounds of less polar compounds for more ease of analysis. New crosslinking techniques which are used today alleviate the degradation problem and have added to the stability of the high polarity columns. The new techniques in making fused silica columns have also decreased the need for the derivatization of drugs for chromatographic analysis. The increased thermal stability and inertness given to crosslinked columns allow for more analysis of drugs without the added step of forming derivatives. The resulting chromatography is reproducible, without loss of selectivity or column capacity (Plotczyk & Larson 1983).

2.6 DERIVATIZATION PROCEDURES

Although technology is enabling the forensic drug chemist to circumvent the need for the use of derivatizing reagents, this is still a popular approach to gas chromatography. The need to form a derivative is seen especially when dealing with the very polar drugs. This problem is encountered more frequently in forensic toxicology, where the metabolic processes of the body have converted drugs to a significantly more polar compound. However, there are drugs of such polarity that also require derivatization in the drug laboratory as well. There are many different commercially available reagents used to derivatize drugs. These various products work to form three typres of derivative compounds, silyl, halogenated and other non-halogenated. Each of these has uses which makes it favored in different circumstances. The end result of each of these, however, is the formation of a compound which is less polar and more volatile than the original compound. This combination of properties allows lower operational temperatures in the GC analysis while keeping adsorption of the compound onto the solid phase minimal, resulting in increased sensitivity and better chromatograhy than the parent compound would have shown under the same conditions.

2.6.1 Silylating agents

This method of derivatization is probably the most popular. Part of this popularity lies in the fact that the resulting compounds are able to react with a large number of drugs of different types, yielding compounds which can easily be detected by using a flame ionization detector. The reagents themselves are highly volatile. The derivatives are unstable in that they can be easily broken down into the original drug in the presence of water. This means that the reagents and solvents used must be kept free of traces of aqueous solvents. The reaction mixtures must also be kept water free. Drugs with free carboxylic acid, phenolic, amide and alcohol groups respond well to different strengths of silylating reagents. Two very polular silylating agents which are also among the stronger of them are bis(trimethylsilyl)trifluoroacetamide (BSTFA) and bis(trimethylsilyl)acetamide (BSA). A typical reaction for a polar compound and a silylating agent is shown below:

$$2\,[RCO_2H] + R_1CON \begin{cases} Si\,(CH_3)_3 \\ SI\,(CH_3)_3 \end{cases} \rightarrow 2\,[RCO2Si\,(CH_3)_3] + R_1CONH_2$$

Target compound — Silylating agent

The stoichiometry of the reaction is easily observed to give two molecules of derivatized drug for each molecule of agent. Once again, owing to the susceptibility of the product to hydrolysis back to the original compound, the greatest care must be taken to keep all reagents and reaction steps free of water. In most cases, part of the reaction mixture is directly injected into the GC without any attempt to isolate the derivative from excess reagents. As one would expect, however, with all of the excess chemicals flowing through the gas chromatograph, the detector of the GC will need cleaning more often than if underivatized drugs are injected (Jack 1984). These reagents have been used on barbiturates, but some scientists have reported variable recoveries of product and recommend that this type of reagent be used for qualitative work rather than quantitative analysis (Street 1971).

2.6.2 Nonhalogenating agents

These agents tend to alkylate the parent drug. There are advantages in that the compounds obtained are relatively stable, making post-derivatization clean-up possible. Again, many classes of drug will respond to this type of agent, and the resulting derivatives are detectable by FID. Most of the reagents used are stable with water, and include such compounds as acetic anhydride which is used in many laboratories. Other agents of this type include diazomethane, tetramethylammonium hydroxide, and, in some cases, methanolic hydrochloric acid. Many of the reactions need a catalyst, and heat to force the reaction to completion. These agents have been used to perform on-column derivatization of barbiturates, though some authors reported apparent breakdown of the barbiturates as exemplified by the appearance of extra peaks in the chromatograms (Pillar & Dilli 1981). A typical nonhalogenating derivatization reaction is shown below.

$$RCOC1 + R_1CH2OH \rightarrow RCO_2CHR_1 + HC1$$

Alkylating Agent — Target compound

2.6.3 Halogenating agents

These agents are used when a compound sensitive to ECD analysis is desired. Like the nonhalogenating agents, the compounds formed are stable, but like the silylating agents, the reactions are very sensitive to the presence of water. Agents used include trifluoroacetic anhydride, other anhydrides and halogenated benzyl compounds. Some halogenated silylating compounds are in use, though the resulting compounds can have very high retention times compared to the TMS counterparts.

One further aspect to consider when using these agents is the possible formation of highly acidic byproducts which upon injection can damage the column (Jack 1984).

2.7 ANALYSIS FOR DRUGS OF FORENSIC INTEREST

2.7.1 Overview

The forensic drug chemist is called upon to analyze many types of samples ranging from commerically produced pharmaceuticals, illicitly manufactured tablets and capsules, powders which are confiscated on the street, to liquids seized in raids on clandestine laboratories. The examination of commercial products is usually straightforward, because the compounds of forensic interest are generally few and in most cases easily extracted from the tablet matrix. As always, there are exceptions which can cause problems, for example pharmaceutical products which contain multiple constituents.

The most common form of sample which enters the forensic drug laboratory apart from standard dosage forms are unknown powders. Whether the powder is in the form of cocaine mixed with other local anesthetic agents, or mixtures of sympathomimetic amines, special considerations must be made before any chromatographic analysis is possible. I will briefly touch on sample pretreatment and subsequently explore groups of compounds and some of the gas chromatographic analytical techniques which have been used in their determination. The groups examined will be cannabinoids, stimulants, barbiturates, benzodiazepines, narcotics and opiates, hallucinogens, and cocaine.

2.7.2 Cannabinoids

Although cannabis is generally identified in most laboratories by use of a color test (modified Duquenois test), and or microscopical examination, there is sometimes the need for some type of additional chromatographic analysis. Gas chromatographic analysis using different temperature programs can give information about batch origin, geographical origin, or simply identification of cannabinoids. The columns most frequently used for the examination of cannabinoids include methylpolysiloxane or phenlymethlypolysiloxane (Novotny *et al.* 1976). Derivatization is not generally necessary for the examination of either the raw plant material or processed products such as hashish or hash oil. Usually some type of organic wash with petroleum ether will yield enough cannabinoids to effect a chromatographic analysis. After redissolving the extract in a solvent such as methanol, the solution can be injected directly onto the gas chromatograph. The cannabinoids cannabidiol (CBD), tetrahydrocannabinol

(THC), and cannabinol (CBN), can be readily separated and preliminary identification achieved with a FID detector, or structural information can be achieved with a mass spectrometer. Variations on this theme were used by Stromberg (1971). He performed various partition experiments by using methanolic extracts of marihuana cigarette smoke residues or hashish. Thermal degradation and the effect of pH on extraction of cannabis components was determined. Under all of his experimental conditions,

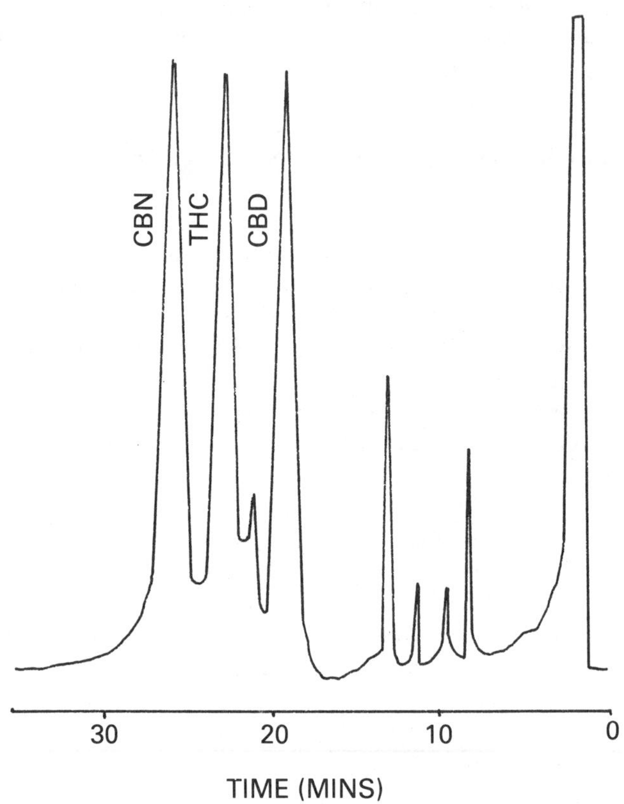

Fig. 2.6. Chromatographic separation of the cannabinoids CBN, THC and CBD (Stromberg 1971).

the CBN, CBD, and THC were easily found (see Fig. 2.6). Variations were noted in the lighter components (Stromberg 1971). Some work has been done with headspace analysis of the cannabis extracts. This type of analysis yields data about more volatile subtances in the cannabis products. However, unless the scientist is interested in comparing the possible common origins of two samples, this will not aid him in the preliminary identification of either THC or the other principal components of cannabis. Hood & Barry (1978) experimented with just such a technique. Headspace analysis was performed on samples of marihuana and hashish to try to find correlations between geographical origin and the relative concentrations of different volatiles. The results were that correlation between geographical origin could not be determined in this way, but could help in the determination of common origin between samples. Further attempts have met with differing degrees of success in the determination of geographical origin of marihuana samples. These generally entail different extraction techniques and subsequent instrumental analysis. An example was the use of cyclohexane to extract polar components from samples followed by a nitromethane wash of the cyclohexane was evaporated to dryness and after reconstitution, analyzed by GC. The results apparently demonstrated geographical as well as regional differentiation in samples (Novotny *et al.*, 1976).

2.7.3 Stimulants

The compounds of particular interest in this class belong to the group of drugs known as phenethylamines. These include amphetamine, methamphetamine benzphetamine and other related compounds which are controlled by statute. A hindrance to analysis are other structurally related compounds which are not controlled by law. These include ephedrine, phenylpropanolamine, and caffeine. The first two are found in many legitimate cold remedies, and in formulations which have been designed to mimic pharmaceutical dosage units of the controlled drugs. Caffeine has been found in clandestine preparations of controlled drugs and in "look alike" preparations. In the last few years, the development of capillary column technology has served to make separation of these compounds easier. Many techniques for the analysis of these compounds include derivatization steps. The compounds themselves are volatile, but generally give rise to poor chromatography. The formation of schiff bases, or conversion to amides, account for many procedures, and are good for FID detection. Halogenated derivatives account for many other processes used and are better suited for ECD detection. Certain of the halogenating agents can be used to help in the detection and separation of optical isomers, a subject which will be covered in greater detail. Silylation, while workable, is used less than other derivatization methods (Jack 1984). The compounds thus formed can be analyzed on the all purpose methylsilicone and phenylmethylsilicone columns. Before the recent improvements in capillary column technology, the phases used in chromatographic analysis of the amine-type stimulants were polar in nature. The AOAC *Official Methods of Analysis*, 1990 edition, still lists a 1% carbowax liquid phase for the analysis of amphetamine. Other phases which have been used include polyamide and Apiezon types (Bastos & Hoffman 1974).

The SE-30 and OV-1 liquid phases used in the WCOT capillary columns have proven useful in the chromatography of phenethylamines. The use of rentention indices as seen in Ardrey & Moffat (1981) and Moffat (1975), will allow the analyst to determine that these compounds can be differentiated by using simple techniques without the added derivatization step. The extraction procedure necessary in the sample clean-up phase can be as simple as dissolving the powder from a tablet or capsule in 0.1N NaOH and extracting with hexane. Conversion to the HC1 salt by bubbling HC1 fumes through the hexane can be beneficial since the bases are generally volatile and can be lost during the evaporation of the hexane. The HC1 salts are less likely to be lost, yet volatile enough for injection onto the GC.

As previously mentioned, some silylating reagents can be used to aid in the determination of optical isomers of the phenethylamines. One which has been investigated is N-trifluoroacetyl-L-prolyl chloride (TPC). This reagent reacts with the l-isomer to form a l-proline derivative which can be chromatographically differentiated from the L-proline D-isomer (Fig. 2.7). The column used in this study was a 1% carbowax 20M phase on 100-120 mesh Gas Chrom Q. A later study used methamphetamine, 1-TPC, with two different columns, SP-2100 and Chirasil-Val. Both columns separated the isomers of the methamphetamine from one another as well as from caffeine, a common diluent (Liu *et al.* 1982).

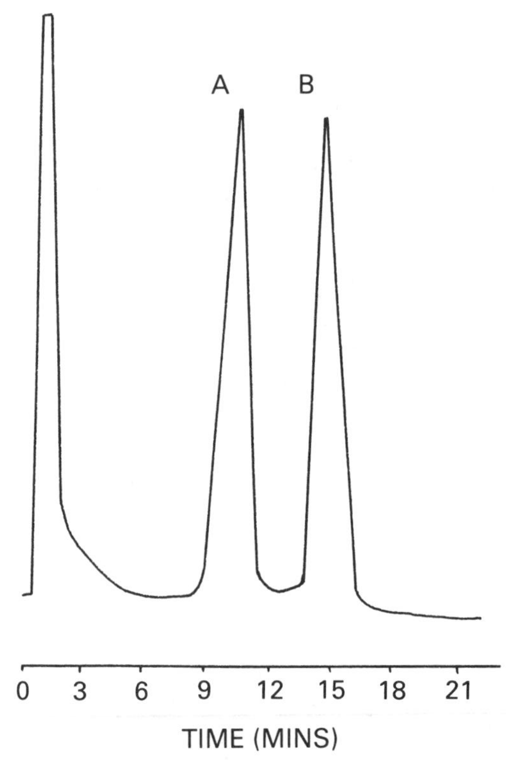

Fig. 2.7. Chromatographic separation of D and L amphetamine (Wells 1970) (A) l-proline, L-amphetamine. (B) l-proline, D-amphetamine.

The use of chiral differentiating agents was seen as a necessity because of the way laws were written in some jurisdictions. Differences in pharmacological activity between D and L optical isomers of controlled substances provided a point of contention between the defense and the prosecution scientist. Further studies using chiral agents demonstrated separation of most phenethylamines, their metabolites and their optical isomers (Jack 1984). Like other drugs of abuse, this class of compound has seen an increase and decrease in its popularity over recent years. At present we seem to be experiencing an upswing in the use of amphetamine like drugs in the United States, as "ice" is hitting the drug market. Known to forensic scientists as methamphetamine free base, its determination by using any of the above methods can be assured. The capillary columns and temperature programs in place in most laboratories today can easily handle this new form of a well known drug. Using the mass spectrometer as a detector does require the forensic scientist to remember that the underivatized phenethylamines are relatively low in molecular weight and give spectra which are in some cases very similar. Care must therefore be taken in identification by this method.

2.7.4 Barbiturates

The barbiturates have been previously mentioned under the discussion of derivatization. While chromatographic separation of barbiturates was first described in the 1960s, the methods have changed somewhat. The first determinations were made by using pyrolysis products (Liu *et al.* 1982). Since that time, work has been performed using determination by dual column analysis (Janak & Hrivnac 1960), derivatization by many methods and reagents, neutralization of active sites on the column before barbiturate injection, and finally, direct injection of the barbiturate itself. Progress to this point had been achieved by 1968, when a 10% SE-30 packed column was used to separate the six barbiturates most commonly used in pharmaceutical preparations (Kazyak & Knoblock 1963). It was found that although the separation desired could be achieved, the peak shape was the factor which hindered anlysis. Chromomatographically, the barbiturates have a tendency to tail, leading to a perceived lack of separation. Experiments designed to eliminate tailing including preinjection of formic acid to occupy active sites on the column (Allen 1968). In addition, many mixed phase columns were tried with moderate success in the separation of barbiturates.

The derivatization techniques for a long time were the most often used until the advent of the modern capillary columns. As previously discussed, silylation and alkylation could be used in conjunction ith an FID detector and halogenation with ECD. The reagents used were many, including BSTFA for silylation, TMAH for methylated compounds and pentaflourobenzyl bromide (PFBB) for halogenation. A study of alkylation seemed to show that butylation accomplished greater separation of barbiturates than methylation or ethylation (Welton 1970). These were all off-column techniques. On-column derivatization was attempted in the late 1960s and the 70s. The use of TMS derivatives was tried, but as the reaction products are relatively unstable compounds, there were resultant peaks which interefered with the chromatographic analysis (Greedy 1974). Alkyl derivatives enjoyed greater success, especially when trimethylphenylammonium hydroxide (TMPAH) was used (Grove & Toseland 1970).

Both electron impact (EI) and chemical ionization (CI) mass spectrometry have been used for the detection of barbiturates. The need for both EI and CI was due to the fact that some barbiturates gave very similar spectra under various conditions (Pillai & Dilli 1981). Many of these studies were developed for the determination of barbiturates in body fluids, and as such are more complex than typical forensic drugs analysis because of concerns over interference from the biological matrix. Straight analysis of barbiturates in pharmaceutical preparations had been accomplished by 1970, using 3% OV-17 on a glass packed column (Kananen 1972). A 1% SE-30 column had also been reported to give an adequate separation of these compounds

(Alber 1969). Again, with the improved capillary technology, separation of even complex mixtures of barbiturates became easier (Ardrey & Moffat 1981).

The sample preparation for barbiturates can be accomplished as follows. Most forensic chemists use a simple acid/organic solvent extraction, or even a direct extraction into ethanol. This extract can be dried and subsequently injected onto the GC. Use of the methyl- or phenylmethylpolysiloxane columns has been shown to give quite adequate separation, and diligent use of the retention times or MS data can often achieve confirmatory analysis of these compounds.

2.7.5 Benzodiazepines

The benzodiazepines were introduced in the 1960s and have four main medical uses. They are used as muscle relaxants, antianxiolytics, anticonvulsants, and sleep inducers. From a forensic point of view, the drug contents in each of the varieties of benzodiazepine dosage units are more than adequate for analysis. While extensive work with GC has been done, there seems to be a tendency toward benzodiazepine analysis by HPLC. Nonetheless, packed columns have been used for the examination of benzodiazepines, using OV-17, 3% on Gas Chrom Q at 280°C (McNair & Bonelli 1987). Other work using the same column, but with an electron capture detector (ECD), has been described, and ten compounds along with metabolites were separated (DeMeijer 1973). Clark's *Isolation and Identification of Drugs* lists retention times for many of these compounds. OV-1 and SE-30 have been used and rention indices listed in previously referred to studies (Ardrey & Moffat 1981, Moffat 1975). Plotczyk & Larson (1983) studied the behavior of drugs on fused silica columns with the 5% phenyl methylsilicone (DB-5, SE-52) and separated, in serum, many drugs, of which two were benzodiazepines, demonstrating that not only can separation be attained between the class, but also from other drugs. One complaint used to be that the retention times for some of the benzodiazepines were very long, but with the recent column and stationary phase improvements, higher running temperatures have cut down retention times.

Extractions from tablets can be attained via solvent extractions for subsequent analysis. When the FID is used as a detector, these drugs give adequate response even into the nanogram levels, so the forensic chemist will have no problem in detection. If MS is used, enough spectral information can be gained for confirmation. One thing to watch out for, however, is breakdown. A prime example is chlordiazepoxide which has a tendency to break down to desmethyldiazepam. With all of the new drugs that are being developed in this class, we are certain to encounter further benzodiazepines with similar properties.

2.7.6 Narcotics and opiates

This class of drug, exemplified by heroin, morphine, meperidine, and codeine, has one aspect which makes chromatographic analysis of this class somewhat easier than most. That is the great differences in polarity between members of this class. Morphine and codeine are the most abused natural narcotics, with heroin, meperidine, hydromorphone and hydrocodone following as synthetic and semi-synthetic narcotics. These and other opiates have been separated on the SE-30 stationary phase (Moffat

1975), as well as OV-17 (Ardrey & Moffat 1981). Analysis for these drugs, is however, not without problems. Morphine, because of its amphoteric nature, is not only difficult to extract, but it tends to give somewhat nonreproducible quantitation results in its underivatized form. It has been reported that if the TMS derivative is made, the variability in quantitation results obtained can be reduced (de Silva *et al.* 1976). The derivatized form does have the disadvantage of a longer retention time, however, and suffers the other problems associated with derivatized drugs described earlier. Work has been performed on isomers of heroin to determine the ability of various instruments to differentiate them. Of interest to us is the efficiency of GC to distinguish these isomers. SE-30 was the liquid phase, and the compounds of interest were derivatized with TMAH. The resulting GC analysis was unable to differentiate all isomers when a packed column was used. However, when a capillary column was substituted, separation was accomplished (Prager *et al.* 1979). These compounds, which were structurally very similar, again demonstrated the ability of the polydimethylsiloxane capillary columns to accomplish separations which packed columns could not approach in the working laboratory.

The fentanyl compounds came to the front when "china white" was isolated on the west coast. This analog of fentanyl was among the first designer drugs. Since the appearance of this compound, work has been done to synthesize and differentiate different analogs of the parent drug. Once again, when scrutinized by GC, 30 related compounds in this group were differentiated. The column was again a capillary column, but the liquid phase was SE-54, and a temperature program was used. The analogs could, under these conditions, give some structural data based on retention times and the functional groups which were present on the molecule (Klahr *et al.* 1989). The more varied structures of the narcotics gave a wider range of properties which can be exploited to allow easy GC separation. This is in contrast to the barbiturates, amphetamines and benzodiazephines which have very closely related class structures and, therefore, similar chemical properties.

2.7.7 Hallucinogens

This class of drug is somewhat similar to the narcotics in that there is a very wide range of structures and properties represented. Some of the most common of this class are phencyclidine (PCP), lysergic acid diethylamide (LSD), methylenedioxyamphetimine (MDA) and its analogues, psilocybin, psilocyn, and mescaline. There have been other hallucinogenic drugs encountered over the years although most of these have come and gone. Although these compounds vary in structure, and would allow us to think that separation should be relatively straightforward, some are problematic. Some cannot be chromatographed unless derivatized, while others are difficult to extract into organic solvent. LSD, for example, was once very difficult to chromatograph since it suffered thermal degradation. Many extractions cleave the phosphoryl group from psilocybin thus converting it to psilocin. Further problems are that some of these drugs have isomers which can complicated anlysis. LSD, for example, has iso-LSD and the methyl-propyl analog LAMPA, which both have atomic weights of 323 amu and result in similar mass spectra. Chromatographic data can aid greatly in the final identification of compounds such as these. A study was made to differentiate several of the more common hallucinogens in 1969. The drugs compared were psilocybin, LSD, DMT (*N-N*-dimethyltryptamine), mescaline, and ibogaine. All compounds were separated when they were converted to TMS derivatives. The column used was an SE-52 liquid phase, with a temperature program (Cooper *et al.* 1986). LSD was the subject of another study where SE-30 was coated on glass beads in a packed column. This procedure used neither derivatization nor temperature programming, and successfully chromatographed LSD, although no attempt was made to differentiate between it and the iso compound or LAMPA (Lerner & Katsiaficas 1969). This separation was accomplished in a later study. The column used was a 25 m fused silica BP-1, 25 m. The GC was used isothermally, and no derivatization was necessary. LSD, LAMPA, and other related compounds were successfully separated (Fig. 14) (Japp *et al.* 1987a,b).

Phencyclidine (PCP) has been abused for many years in the United States. Sold under various names in an attempt to hide its identity from a bad reputation, PCP has been found mixed with plant materials including cannabis, painted on cigarettes, and in powder and liquid forms. Since the drug is no longer made legitimately, the only sources are clandestine laboratories. The quality control is, needless to say, not up to the standards of a pharmaceutical company and many impurities can be found. This allows us to not only identify the PCP by GC, but also to determine the synthetic pathway used to prepare the drug. SE-30, OV-17 and other liquid phases on packed columns were used in one study to determine the chromatographic properties of some of these compounds. Separation was accomplished quite easily on most phases with SE-30 offering the best performance (Katz *et al.* 1967). PCP is being used in some laboratories to evaluate the formation of active sites on columns. Apparently it is easily captured by these active sites when they form, resulting in decreased elution from the column.

The 3,4-methylenedioxyamphetamine analogs have been seen as designer drugs in recent years. Their characterization has been studied on glass packed and capillary columns of different liquid phases. Separation can be accomplished without derivatization, and quantitation is possible if care is taken to increase the oven temperature after elution of the compounds. This increase prevents deterioration of peak shape and changing retention times. Capillary columns again seem to be more appropriate for the chromatographic examination of these compounds (Cone *et al.* 1979)

2.7.8 Cocaine

We have dealt with cocaine separately, not only because it has been termed, amongst other things, a narcotic and a stimulant, but because it is the most common drug of abuse in the United States. Cocaine is a local anesthetic and in illicit preparations is often found in combination with other local anesthetics such as procaine, lidocaine and tetracaine. Tests on many of the local anesthetics have been described, using an isothermal run with a glass column and OV-17 on Gas Chrom Q. Separation was accomplished in reasonable run times (Dal Cason 1989). Previously, the AOAC had adopted a gas chromatographic procedure which utilized a 3% OV-1 column for preliminary identification and quantitation of cocaine from tablets and powders (Clark 1986). Neither of these used derivatization. Procedures have been used to determine other constituents of the coca leaf which have been extracted along with the cocaine. These have used a TMS derivative and 3% OV-25 on Gas Chrom Q, and 10% OV-101 on Chromasorb W-HP (Baker & Gough 1979). From these examples, although the diluents can be closely related, the required separation can be accomplished.

2.8 SUMMARY

We have seen that most drugs of abuse have the potential for gas chromatographic analysis. The problems we have faced are diluents which are structurally related to the drug of interest, lability of the compound, and other abused drugs of the same class which respond in a similar manner. These problems have been overcome in the most part with the introduction of capillary columns. This technology has made our life easier not only because of the higher column efficiency which allows greater separating power, but also because generally lower concentration of the drug is necessary for analysis. Both licit and illicit drugs are being introduced which are more potent and therefore found in less concentration in a dosage unit. This is a trend which will undoubtedly continue. It is hoped that forensic drug analysis together with instrument manufacturers will respond to this challenge with the development of more sensitive and specific analytical techniques.

It has been said that there is no ideal column for the analysis of all of the drugs of abuse. Even with the capillary columns, this is still true. However, with todays columns, I do feel that two capillary columns which have even slightly different polarity characteristics can be used to solve most of the chromatographic problems which forensic laboratories may face with drugs of abuse.

REFERENCES

Alber, L.L. (1969), *Journal of the AOAC*, **52** (6) 1295.

Allen, J.L. (1968), *Journal of the AOAC*, **51** (3) 619.

Ambre, J., Belknap, S.M., Nelson, J., Ruo, T.I., Shin, S.-G. & Atkinson, A.J. (1988). *Clin. Pharmacol. and Therapeutics*, **44** (1) 1.

Ardrey, R.E. & Moffat, A.C. (1981), *Journal of Chromatography*, **220** 195.

Bagasra, O. & Forman, C. (1989), *J. Immunol.*, **77** 289.

Baker, P.B. & Gough, T.A. (1979), *Journal of Forensic Sciences*, **24** (4) 847.

Barnfield, C., Buns, S., Byrom, D. & Kemmenoe, A. (1988), *Forensic Science Inernational* **39** 107.

Bastos, M.L., Hoffman, M.L. & Hoffman, D.B. (1974), *Journal of Chromatographic Science*, **12** 269.

Both (1990).

Chapman, D.I. (1986), Infra-red spectrophotometry. In Clarke's *Isolation and identification of drugs*, 2nd ed., Moffat, A.C., ed. 237–245, London, The Pharmaceutical Press.

Chasnoff, I., Bussey, M.E., Savich, R. & Stack, C.M. (1986), *J. Pediatrics*, **108** (3) 456.

Clark, C.C. (1978), *Journal of the AOAC*, **61** (3) 683.

Clark, C.C. (1986), *Journal of the AOAC*, **69** (5) 814.
Clark, E.G.C. (1955), *Bull. on Narcotics*, **7** (3–4) 33.
Clark, E.G.C. (1961), *Bull. on Narcotics*, **13** (4) 27.
Clark, E.G.C. (1962), *Methods of Forensic Science*, **1** 9 8.
Cooper, D., Jacob, M. & Allen, A. (1986), *Journal of Forensic Sciences*, **31** (2) 511.
Cone, E.J., Darwin, W.D., Yousefnejad, D. & Buchwald, W.F. (1979), *Journal of Chromatography*, **177** 149.
Dal Cason, T.A. (1989), *Journal of Forensic Sciences*, **34** (4) 928.
DeMeijer (1973).
de Silva, J.A.F., Bekersky, I., Puglisi, C.V., Brooks, M.A. & Weinfeld, R.E. (1976), *Analytical Chemistry*, **48** (1) 10.
Dixon, S.D. & Bejar, R. (1989), *J. Pediatrics*, **115** (5) 770.
Fischman, M. (1988), *J. Clin. Psychiatry*, **49** (2 supp) 7.
Fulton, C.C. (1962), *Microchemical Journal*, **6** 51.
Fulton, C.C. (1969), Selected color test reagents. In *Modern microcrystal tests for drugs*, Fulton, ed. 335–339, New York, Wiley-Interscience.
Galvin, F. (1988), *J. Clin. Psychiatry*, **49** (2 supp) 11.
Giannini, A.J. (1989), Phencylidine. In *Drugs of abuse*, Giannini & Slaby, eds. 145–159, New Jersey, Medical Economic Books.
Giannini, A.J., Slaby, A.E. & Giannini, M.L. (1985), *Handbook of overdose and detoxifiction emergencies*, 3rd ed., Medical Examination Publishing Co., New York.
Gold, M.S. & Giannini, A.J. (1987), Cocaine and cocaine addiction. In *Drugs of abuse*, Giannini & Slaby, eds. 83–95, New Jersey, Medical Economic Books.
Gold, M.S. (1989), Opiates. In *Drugs of abuse*, Giannini & Slaby, eds. 127–144, New Jersey, Medical Economic Books.
Greely, R.H. (1974), *J. Chromatog.*, **88** 229.
Grove, & Toseland, P.A. (1970), *Clin. Chimica Acta*, **29** 213.
Hindmarsh, K.W. & Opheim, E. (1990), *The Int. J. of the Addictions*, **25** (3) 301.
Hoffman, F. (1983a), CNS stimulants. In *A handbook on drug and alcohol abuse: the biomedical aspects*, House, ed. 227–247, Oxford, New York, Oxford University Press.
Hoffman, F. (1983b), Hallucinogens, LSD, phencyclidine and other agents having similar effects. In *A handbook on drug and alcohol abuse: the biomedical aspects*, House, ed. 170–173, Oxford, New York, Oxford University Press.
Hoffman, F. (1983c), Narcotic drugs. In *A handbook on drug and alcohol abuse: the biomedical aspects*, House, ed. 68–90, Oxford, New York, Oxford University Press.
Hoffman, F. (1983d), Drug diluents: general aspects of drug abuse. In *A handbook on drug and alcohol abuse: the biomedical aspects*, House, ed. 29–36, Oxford, New York, Oxford University Press.
Hood, L.V.S. & Barry, G.T. (1978), *Journal of Chromatography*, **166** 499.
Jack, D.B., ed. (1984), *Drug analysis by gas chromatography*, Academic Press, Orlando, Florida.
Jackson, J.V. (1986), Forensic toxicology. In Clark's *Isolation and identification of drugs*, 2nd ed., Moffat, A.C., ed. pp. 35–53, London, The Pharmaceutical Press.
Janak, J. & Hrivnac, M. (1960), *J. Chromatog.*, **3** 297.

Japp, M., Gill, R. & Osselton, M.D. (1987a), *Journal of Forensic Sciences*, **32** (4) 933.
Japp, M., Gill, R. & Osselton, M.D. (1987b), *Journal of Forensic Sciences*, **32** (6) 574.
Jatlow, P. (1988), *The Yale Journal of Biology and Medicine*, **61** 105.
Javaid, J.I., Fischman, M. & Schuster, C.R. (1978), *Science*, **202** 227.
Johanson, C. & Fischman, M. (1989), *Pharmacological Reviews*, **41** (1) 3.
Jones, R.T., Benowitz, N. & Bachman, I. (1976), *Ann. N.Y. Acd. Sci.*, **282** 21
Kananen, G. (1972), *J. Chromatographic Sci.*, **10** 283.
Katz, M.A., Tadjer, G. & Aufricht, W.A. (1967), *Journal of Chromatography*, **31** 545.
Katzung, B. ed., (1989a), Drugs of abuse. In *Basic and clinical pharmacology* 9th ed., 383–394, Connecticut/California, Appleton and Lange.
Katzung, B. ed., (1989b), Opioid analgesics and antagonists. In *Basic and clinical pharmacology*, 4th ed., 368–382, Connecticut/California, Appleton and Lange.
Katzung, B. ed., (1989c), Local anesthetics. In *Basic and clinical pharmacology*, 4th ed., 315–322, Connecticut/California, Appleton and Lange.
Kazyak, L. & Knoblock, E. (1963), *Analyt. Chem.*, **35** 1448.
Klahr, A.L., Roehrich, H.G. & Miller, N.S. (1989), Marijuana. In *Drugs of abuse*, Gianni, A.J. & S,aby, A.E. eds, Medical Economoics Books, New Jersey.
Kolodny, R.C., Masters, W.H., Kolodner, R.M. & Toro, G. *New Eng. J. Med.*, **290** (16) (1974).
Kovats, E. (1961), *Z. Anal. Chem.*, **181** 351.
Larsen, A.K., Illinois State Police, *Pers. comm.*
Lerner, M. & Mills, A. (1963), *Bull. on Narcotics*, **15** (1) 37.
Lerner, M. & Katsiaficas, M.D. (1969), *Bulletin on Narcotics*, **21** (1) 47.
Linn, S., Schoenbaum, S.C., Monson, R.R., Rosner, R., Stubblefield, P.C. & Ryan, K.J. (1983), *Am. J. Public Health*, **73** 1161.
Liu, J.H., Ku, W.W., Tsay, J.T., Fitzgerald, M.P. & Kim, S. (1982), *Journal of Forensic Sciences*, **27** (1) 39.
Loper, K.A., (1989), *Med. Toxicol. Adverse Drug Exp.*, **4** 174.
McNair, H.M. & Bonelli, E.J. (1967), *A basic gas chromatography*, Varian Aerograph, Walnut Creek, C.A.
McReynolds, W.O. (1970), *J. Chrom. Sci.*, **8** 865.
Medina, F. (1989), *Journal of Forensic Sciences*, **34** (3) 565.
Meyers, J. *Drug Enforcement Administration*, Chicago. *Pers. comm.*
Miller, N.S. & Gold, M.S. (1989), Amphetamine and its derivatives. In *Drugs of abuse*, Giannini & Slaby, eds. 15–42, New Jersey, Medical Economic Books.
Moffat, A.C. (1975), *Journal of Chromatography*, **113** 69.
Moffat, A.C. (1986a), Monographs; Analytical and toxicological data. In Clarke's *Isolation and identification of drugs*, 2nd ed. Moffat, ed. 484–490, London, The Pharmaceutical Press.
Moffat, A.C. (1986b), Monographs; Analytical and toxicological data. In Clarke's *Isolation and identification of drugs*, 2nd ed. Moffat, ed. 349–350, London, The Pharmaceutical Press.
Moffat, A.C. (1986c), Monographs; Analytical and toxicological data. In Clarke's *Isolation and identification of drugs*, 2nd ed. Moffat, ed. 763–764, London, The Pharmaceutical Press.

Moffat, A.C. (1986d), Monographs; Analytical and toxicological data. In Clarke's *Isolation and identification of drugs*, 2nd ed. Moffat, ed. 874–875, London, The Pharmaceutical Press.
Moffat, A.C. (1986e), Monographs; Analytical and toxicological data. In Clarke's *Isolation and identification of drugs*, 2nd ed. Moffat, ed. 524–525, London, The Pharmaceutical Press.
Moffat, A.C. (1986f), Monographs; Analytical and toxicological data. In Clarke's *Isolation and identification of drugs*, 2nd ed. Moffat, ed. 490–491, London, The Pharmaceutical Press.
Montagne, M. (1990), *The Int. J. of the Addictions*, **25** (5) 557.
Moore, J.M. (1974), *Journal of Chromatography*, **101** 215.
Novotny, M., Lee, M.L., Low, C.E. & Raymond, A. (1976), *Analytical Chemistry*, **48** (1) 24.
Pillai, D.N. & Dilli, S. (1981), *Journal of Chromatography*, **220** 253.
Plotczyk, L.L. & Larson, P. (1983), *Journal of Chromatography*, **257** 211.
Prager, M.J., Harrington, S.M. & Governo, T.F. (1979), *Journal of the AOAC*, **62** (2) 304.
Raverby, M. (1987), *J. Forensic Sciences*, **32** (1) 20.
Raverby, M. & Gorski, A. (1989), *J. Forensic Sciences*, **34** (4) 918.
Resnick, R.B., Kestenbaum, R. & Swartz, L.K. (1978), *Science*, **195** 696–698.
Rohrscheider, L.J. (1966), *J. Chromatog.*, **22** 67.
Schawbell, J.L. (1990), *Emergency medicine clinics of North America*, **8** (3) 595.
Shannon, M. (1988), *Ann. Emerg. Med.*, **17** 1243.
Siegel, R.K. (1982), *J. Psychoactive Drugs*, **14** 271.
Smith, R. & Stewart, J.T. eds. (1981), Separation and purification methods. In *Text book of biopharmaceutic analysis*, 27–49, Philadelphia, Lea and Febiger.
Stevens, H.M. (1986), Color tests. In Clarke's *Isolation and identification of drugs*, 2nd ed. Moffat, A.C., ed. 128–147, London, The Pharmaceutical Press.
Stowell, A. & Wilson, L. (1987), *Journal of Forensic Sciences* **32** (5) 1214.
Street, H.V., (1971), *Clin. Chim, Acta*, **34** (2) 357.
Stromberg, L.E. (1971), *J. Chromatog.*, **63** 391.
Turner, C.E. (1981), *The marijuana control controversy: definition, research perspectives and therapeutic claims.* The American Council for drug Education, Washington D.C.
Van Dyke, C. & Byck, R. (1982), *Scientific American*, **246** (3) 128.
Van Dyke, C., Barash, P.G., Jatlow, P. & Byck, R. (1976), *Science*, **191** 859.
Van Dyke, C., Jatlow, P., Ungerer, J., Barash, P.G. & Byck, R. (1978), *Science*, **200** 211.
Wells, C. (1970), *Journal of the AOAC*, **53** 113.
Welton, B. (1970), *Chromatographia*, **3** 211.
Wetli, C.V. & Wright, R.K. (1979), *JAMA*, **241** (23) 2519.

3

Gas chromatographic applications in forensic toxicology

Christine M. Moore Ph.D. and Ian R. Tebbett Ph.D.
Department of Pharmacodynamics, University of Illinois at Chicago, Box 6998, Chicago, IL 60680, USA

3.1 INTRODUCTION

Forensic toxicology can be broadly defined as the study of toxic substances encountered in the course of judicial investigations. In recent years, the bulk of the work performed in the forensic toxicology laboratory has involved analysis for drugs and can therefore be considered as an extension of the work of the drug chemist. The difference is that the toxicologist is usually asked to identify and quantify drugs or other toxic substances in post mortem samples. In many cases he is also required to interpret this information in order to determine whether or not these compounds were a cause or contributing factor in the death of the victim. There are over 10 000 toxic substances known although only a few of these are encountered in forensic cases such as homicide, suicide or accidental death.

The recent move by companies toward pre-employment and random drug screening of their employees has meant that some laboratories under the direction of a forensic toxicologist are now called upon to examine clinical samples such as blood and urine, for the presence of illicit substances principally cocaine, opiates, cannabinoids, PCP and amphetamines as well as the investigation of the thousands of deaths resulting from overdose of these compounds. In addition the toxicology laboratory performs alcohol determination in blood and urine in cases of driving under the influence (Chapter 4).

As well as drug overdose and accidental deaths, the forensic toxicologist can also confirm or refute the claims of the suspect or victim. For example allegations of doping prior to rape or robbery can be verified by the examination of the appropriate blood or urine samples, and a knowledge of the pharmacokinetics and pharmacology of the drug in question. Blood stains on clothing following violent crimes, can also be examined for the presence of drugs.

Samples submitted to the toxicology laboratory for analysis are usually removed by the pathologist at autopsy. These include blood, urine, brain, kidney and bile. The general approach to the examination of these samples is similar to that for the identification of street drugs. That is, a preliminary test is performed to determine the presence or absence of any drugs. The drug is then usually confirmed and quantified by gas chromatography-mass spectrometry after extraction of the drug from the biological sample. Tissue samples are first homogenized in a blender before extraction of the drugs into an organic solvent in the same way as blood and urine are treated. After determining what, if anything, is present and at what concentration, and by considering the levels of drugs found in the body fluids and tissues, an interpretation can be made as to whether the drug was the cause or a contributing factor in the death of the victim.

There are several different groups of compounds commonly encountered by the forensic toxicologist. Of these, however, drugs of abuse account for the greatest percentage of cases. Gas chromatography is an important tool in the identification and quantification of these compounds.

The ever increasing drug abuse facing the United States and other countries throughout the World is well recognized and has been a major concern of recent government administrations. While most people are aware of the so called 'hard' drugs such as cocaine, heroin and amphetamine, this represents the tip of the iceberg with not only controlled substances but also prescription and even over the counter drugs being open to abuse by certain members of the community. The major classes of drugs which possess an abuse potential can be classified as sedative hypnotics, stimulants and hallucinogens.

The application of gas chromatographic techniques to the determination of drugs in biological fluids is by no means new. Gas chromatography has been the chosen technique of forensic toxicologists for many years mainly because of its versatility, high resolution and easy interfacing with mass spectrometry. Absolute identification of a drug and/or its metabolites, such as that provided by mass spectrometry, is required by the legal profession in the United States, and gas chromatographic methods are much easier to interface with mass spectrometers than liquid chromatographic systems. Drug levels encountered in forensic toxicology are usually higher than those found in clinical samples, but the techniques used are the same.

3.2 DRUG SCREENING

Drug screening procedures have been developed which use different GC conditions in terms of columns, temperature programs and detectors. Drugs are among the more active solutes determined by GC, and derivatization is normally required for their analysis on packed columns. The increased inertness of the fused silica open tubular (or capillary) column may allow direct determination of the underivatized drug in some cases. Derivatization of biological extracts prior to analysis by GC is required to impart thermal stability, to give additional evidence of identity or to improve gas chromatographic characteristics. This is because many drugs have a high molecular weight and contain relatively polar substituent groups which contribute to interactions with the solid support. For identification purposes, the indication that specific functional groups are present by the successful formation of the required derivatives also provides valuable information.

The chemical properties of the drug under study determine the type of derivative to be used. In the case of drug screening, a general derivatization procedure is often employed. The most commonly used procedure involves adding bis [trimethylsilyl] trifluoroacetamide (BSTFA) (50 μl) with 1% trimethylchlorosilane (TMCS) to the evaporated extract and heating to 90°C for one hour prior to analysis. This procedure produces the trimethylsilyl derivative of the drug. Gibb (pers. comm.) uses this procedure for a forensic drug screen of 430 compounds including 140 parent drugs, metabolites and derivatives.

3.3 DERIVATIZATION PROCEDURES

There are various derivatization procedures which can be adopted depending on the drug or metabolite to be determined.

3.3.1 Silylation

Most alcohols and phenols will form silyl ethers when treated with trimethylsilyl (TMS) reagents such as hexamethyldisilizane (HMDS), trimethylchlorosilane (TMCS), bis (trimethylsilyl) acetamide (BSA) or bis (trimethylsilyl) trifluoroacetamide (BSTFA) at room temperature. Some compounds with sterically hindered hydroxy groups may require heating. Dimethylsilyl derivatives can also be formed by using dimethylchlorosilane (DMCS), tetramethyldisilazane (TMDS), bis (dimethylsilyl) acetamide (BDSA) or chloromethyl dimethylchlorosilane (CMDMCS), but these have lower boiling points and are less thermally stable than their trimethylsilyl counterparts. In all cases, the reagent is used to replace an active proton, rendering the derivatized compound less reactive then the original compound, resulting in a decreased tendency for thermal degradation or hydrogen bonding. Successfully derivatized compounds include sugars, amines, thiols and steroids.

3.3.2 Acetylation

As in silylation, acetylation is used to replace an active proton rendering a compound less polar. Primary and secondary amines can be acetylated readily by using acid anhydrides and heat. Acetic anhydride, propionic anhydride and trifluoroacetic anhydrides are the reagents which are most commonly used. To take maximum advantage of sensitive electron capture detection, heptafluorobutyric anhydride has been used. A typical reaction may be carried out at room temperature for a few hours, but normally temperatures ranging from 50 to 150°C are used with a reaction time of between 15 minutes and 2 hours. Dopamine and epinephrine are often determined as trifluoroacetate derivatives.

3.3.3 Methylation

Methylation derivatization can take place either before analysis or 'on-column'. The on-column method is particularly useful for barbiturates (which do not form stable TMS derivatives) and some carboxylic acids which can simply be mixed with 0.2 M trimethylanilinium hydroxide in methanol prior to injection. The temperature of the column initiates and completes the derivatization process. Methylation can also be achieved by using HCl in methanol or diazomethane. Diazomethane is toxic and potentially explosive, but the reaction time is rapid.

Derivatization, then, is a necessary step in the analysis of many drugs of forensic interest. As each group of drugs is addressed, so specific derivatization procedures will be reported.

3.4 GAS CHROMATOGRAPHY COLUMNS

Over the last twenty years, recommended gas chromatographic systems for drug screening from biological fluids have made use of packed columns, particularly nonpolar phases such as SE-30, OV-1 or OV-101 which are dimethylsilicone polymers. These phases are considered to be the most appropriate for the elution of the compounds of interest. General screening procedures for thermally stable drugs are usually variations on, or modifications of, those described by Ardrey & Moffat (1981). Their method recommends 2.5% SE-30 packing on 80-100 mesh Chromosorb G (acid-washed and dimethylchlorosilane treated). The dimensions of the glass column are 2 m × 4 mm, carrier gas is nitrogen (45 ml/min) and the column temperature between 100 and 300°C. For more polar compounds, Carbowax and OV-17 which are phenylmethyl silicone polymers are chosen. Despite certain disadvantages of packed columns, such as adsorption effects and the lack of sensitivity, they have been favoured for a long time owing to their robustness and inter-laboratory reproducibility.

More recently, the advent of open tubular or capillary column gas chromatography has offered an alternative to packed columns. Because of its much higher separation efficiency, resolution and sensitivity, capillary gas chromatography appeared to be of greater value to the forensic toxicologist. Initial problems with reproducibility of results were observed, but better manufacturing procedures have improved on this. Capillary columns with an internal diameter of less than 0.25 mm allow faster analyses than packed columns but they impose additional demands on the equipment. Because of their small diameters, gas flow volumes and load sample capacities are restricted and fairly specialized injection systems such as split/splitless injectors are needed. Splitless injection, however, is not a solution in itself, since low concentrations of drug may be undetected. Further, with narrow bore columns, an extra supply of make-up gas at the detector end of the column is often necessary to reduce band broadening and allow optimization of the detector.

Drug screening procedures which recommend capillary column systems are widespread (Eklund *et al.* 1983, Soo & Bergert 1986, Manca *et al.* 1989, Cox *et al.* 1989). The International Olympic Committee Medical Commission (IOC-MC) which is responsible for curbing the abuse of doping agents in athletics, conducted its first drug screening tests in the 1968 Olympic Games in Grenoble. The drug screening procedures used at the most recent Olympics event in Seoul, Korea, in 1988 are arguably the best in the World. Systems used incorporate capillary gas chromatography with flame ionization, nitrogen-phosphorus and mass spectrometric detection (Park *et al.* 1990). The tests cover about 100 different drugs and another 400 as metabolites in addition to pharmacologically related substances.

Stimulants (amphetamine, xanthines, cocaine, etc.), narcotic analgesics (methadone, morphine, dipipanone, etc.), and beta-blockers (atenolol, propanolol, etc.), were determined from urine by using a crosslinked fused silica SE-54 or SE-30 capillary column (0.2 mm i.d.; 0.33 μm film thickness) linked to a N-FID or an MSD detector. Variations in column length and film thickness were incorporated when steroid analysis was required (see later).

For forensic purposes, Mule & Casella (1988a) describe the confirmation in urine of marijuana, cocaine, morphine, codeine, amphetamine and phencyclidine by GC/MS. They incorporate a fused silica column (12.5 m $\times$ 0.2 mm i.d.) consisting of crosslinked dimethylsilicone with helium as the carrier gas. The temperature of the column varied with the analyte and the injection was splitless.

The introduction of wide-bore capillary columns has gone a long way toward improving on the drawbacks of both capillary and packed columns. Wide-bore capillary columns have an internal diameter of approximately 0.5 mm, although this estimate can vary. The smallest i.d. which allows direct on-column injection with a standard syringe is 0.53 mm. A thin film coating (0.1 μm) provides a lower sample capacity but shorter retention times since it is less resistant to mass transfer. Wide-bore columns provide increased column lifetime and better reproducibility than their competitors. They offer less resistance to gas flow resulting in minimal pressure drop. Long columns are therefore feasible, offering an increased number of theoretical plates over the packed columns. They can accept large gas flows and are instrumentally compatible with packed column systems, a distinct advantage over capillary columns. Wide-bore columns do not contain support so sample loss due to adsorption or degradation is eliminated.

Caldwell & Challenger (1989) compare the performance of wide-bore capillary columns to packed columns for drug screening. Their procedure successfully identified a wide range of drugs including those of forensic interest, for example amphetamine, methamphetamine and cocaine as well as the polar drugs trimethoprim and quinine. The chromatographic system consisted of a model 5890 gas chromatograph (Hewlett Packard) and a capillary column split-splitless inlet system. A 25 m HP-5 wide-bore (0.32 mm i.d.) thick film (0.52 μm) crosslinked fused silica column was used linked to a nitrogen-phosphorus detector. Helium was used as the carrier gas with nitrogen as the make-up gas. The injector was heated to 250°C and the detector to 300°C. During the analysis, the column temperature was programmed from 90°C for 0.5 minute, raised to 250°C for 0.1 minutes (40°C per minute), then raised again at 5°C per minute to 310°C. The injector was used in the splitless mode so that the entire injection was passed onto the column before activation of the inlet purge at 0.4 minutes.

They conclude that the retention data collected by using this system is reproducible with the exception of polar drugs which tend to give tailing peaks. However, this was also true when packed column systems were used. The capillary system required far less maintenance and was more reproducible and sensitive than packed column methods previously used. The limit of detection achieved for most drugs was 0.25 mg/l).

Chen *et al.* (1990) also use a fused silica wide-bore capillary column with flame ionization detection for screening drugs extracted from plasma. The column was 25QC5/BP1 (25 m × 0.53 mm i.d.; 3.0 μm film thickness). The temperature program ran from 80°C for 2 minutes to 215°C at 20°C/minute, to 285°C at 5°C/minute. It remained at 285°C for 2 minutes. The injector and detector temperatures were 275 and 310°C respectively. Helium was again used as the carrier gas and the injection port was in the splitless mode. The drugs chosen in Chen's study were selected so as to represent various characteristics and classes as well as to cover a relatively wide range of GC retention indices.

In general, screening procedures cover parent drugs and major urinary metabolites. Lillsunde & Korte (1991) report a screening procedure covering 300 substances including all potentially abused drugs and their metabolites. However, special procedures were necessary for the confirmation of buprenorphine, cannabinoids, cocaine, LSD, morphine, phencyclidine, halogenated hydrocarbons, acetaminophen and alcohol. They used a capillary column (25 m × 0.3 mm i.d.) of fused silica crosslinked with 5% phenylmethyl silicone SE54 with a film thickness of 0.17 μm; or a 1 m × 2 mm i.d. column packed with 2% SP-2110/1% SP-2510 on 100/120 Supelcoport. For the capillary column the operating temperatures were between 100 and 300°C; for the packed column between 150 and 240°C.

The examples given are general screens. The development of GC procedures for individual drugs, their metabolites or structurally related groups of drugs of forensic interest are continually being reported.

3.5 SEDATIVE HYPNOTICS

Opium has been used as a sedative since ancient Greek times. It is a natural product obtained from the dried latex of the opium poppy *Papaver somniferum.* Crude opium is a source rich in the narcotic analgesics codeine and morphine and it has been long known that inhalation of the vapors produced by heating opium will cause sedation and a long peaceful sleep. Unfortunately, addiction to the drug develops rapidly. Opium is still occasionally found as a street drug; however, the extensive social and medical problems associated with opiate addiction today followed two scientific discoveries: the isolation of morphine and codeine for analgesia, cough suppression and dysentery, and the invention of the hypodermic syringe. By the year 1900 opiate abuse had reached the United States, with the middle class being the largest single group of addicts. Opiate addiction became recognized as a problem and nonaddictive substances were sought as substitute analgesics. Nowadays, illicit opiate abuse is largely associated with heroin. This drug is produced semi-synthetically by the acetylation of morphine and is also more addictive. Street heroin in the United States is generally of a particularly low quality typically being only 1–2% pure, the remainder consisting of other drugs and diluents such as strychnine, quinine, sugar, starch and even household cleaners such as Ajax. This sample would be dissolved in water or in lemon juice or vinegar, the acidity being considered to give a better dissolution of the drug. The solution is then injected directly into an arm or leg vein or under the tongue into the buccal cavity. This practice results in massive ulceration of the blood vessels, and toxic effects are associated as much with the diluents as with the heroin itself. Heroin may also be heated on aluminium foil and the vapors inhaled. This method of administration is by no means less addictive than intravenous injection.

Tolerance to opiates develops quickly. This is not due to an increase in metabolism, but more to deactivation of the naturally occurring opiate receptors. On withdrawal of the opiate, the receptors become supersensitive and the system becomes overactive resulting in withdrawal symptoms.

The treatment of opiate withdrawal symptoms usually involves the use of an orally administered synthetic opiate such as methadone. Treatment is based on the assumption that cellular changes occur in the addict who then requires a maintenance opiate. Methadone therefore merely substitutes for the heroin. It is however longer acting which reduces the euphoria associated with the heroin, and also gets the addict away from the idea of injecting drugs intravenously.

Opiate antagonists occupy the opiate receptors in the central nervous system, thereby preventing narcotics from exerting an effect. Some of the earlier antagonists such as nalorphine also had agonist actions and were therefore open to abuse. Similarly, buprenorphine was marketed as an opiate antagonist with no abuse potential. This compound which is available as a subcutaneous injectable and as sublingual tablets has however been subject to abuse by opiate addicts.

Opiate users will often take sedative/hypnotics in addition to heroin. This is also the case with cocaine addicts and alcoholics. The short acting barbiturates such as secobarbitone or pentabarbitone are preferred to long acting agents such as phenobarbitone. Other commonly abused sedative hypnotics are meprobamate, methaqualone and the benzodiazepines. Abuse addiction and tolerance develop with all members of this class. The historical development of the sedative hypnotics has seen the introduction of one sedative for another, each drug having its own abuse potential. Bromides were the first sedatives after alcohol to be marketed for their sedative effects in 1826. Barbiturates followed in 1900, then chloral hydrate in 1930, meprobamate in 1955 and the benzodiazepines in 1961. The truth is that it is impossible to treat anxiety, depression and insomnia without the risk of addiction. The nature of the symptoms themselves and the fact that these drugs act directly on the CNS, may be such as to lead to abuse and addiction. Tolerance and dependence regularly accompany frequent drug use and are natural adaptations of the nervous system to the presence of foreign chemicals. The sedative hypnotics depress the activity of all excitable tissue, particularly nerve cells. In large doses the drugs can suppress function in cardiovascular activity. Early effects are loss of attention and concentration, impaired short term memory, euphoria, intoxication and lack of coordination.

Various drugs act synergistically with sedative hypnotics to produce dangerous depression of respiration and ultimately cardiac failure. Ethanol, antihistamines, isoniazid, methylphenidate and monoamine oxidase inhibitors can all increase CNS depressant activity of the sedative hypnotics. In addition the metabolism of some drugs, for example corticosteroids, oral contraceptives digoxin and beta-blockers, can be enhanced as a result of the induction of microsomal enzymes by the sedatives.

Other sedative hypnotics which are open to abuse are largely drugs with anticholinergic or antihistamine activity. In sufficient doses both groups of drugs are capable of causing drowsiness, euphoria, fatigue and hallucinations. Drugs which block muscarinic cholinergic receptors in the CNS are specifically used for the treatment of Parkinson's disease. However, many antihistamines, antipsychotic and antidepressant drugs all possess varying degrees of anticholinergic activity. These include a number of prescription drugs such as; promazine and prochlorperazine as well as over the counter cough and cold remedies which are also abused because of the active ingredient diphenhydramine.

3.6 OPIATES

It is common in forensic urine analysis to test for morphine as a means of identifying opiate or heroin abuse. However, since morphine is also a metabolite of the legally available drugs codeine and ethylmorphine, no firm conclusions can be drawn from the determination of morphine itself in body fluids. Indeed, morphine is present in many prescription preparations for the treatment of pain and cough suppression.

Heroin is first metabolized in the body to 6-acetyl morphine, before being totally converted to morphine. It has therefore been suggested that the determination of the 6-acetyl morphine in urine has much more value as an indicator of heroin abuse (Kintz *et al.* 1987) even though its half life is relatively short (0.6 hours) (Cone *et al.* 1990). Levels of 6-acetyl morphine and free morphine have been reported as being as low as 24 ng/ml in the urine of opiate users (Mule & Cassella 1988b).

Gas chromatographic methods for the determination of 6-acetyl morphine in urine have been reported (Fehn & Megges 1985, Christopherson *et al.* 1987). Paul *et al.* (1988) derivatized the 6-acetyl morphine to its propionyl ester by using propionic anhydride and dry pyridine. Chromatography was carried out on a capillary column (15 m × 0.25 mm i.d. DB-5) at 130 to 250°C (rate 25°C/minute). The reported limit of detection was 810 pg/ml.

Bowie & Kirkpatrick (1989) describe a procedure for the simultaneous determination of morphine, codeine and 6-acetyl morphine in urine by using GC-MS. Their method employs deuterated acetic anhydride to acetylate free hydroxyl groups on all of the opiates. As a result, the same extraction and determination procedures can be used for all the analytes. They used a fused silica capillary column (12.5 m × 0.2 mm i.d.) crosslinked with dimethyl silicone. The detection limit was less than 10 ng/ml for all analytes.

A considerable amount of ingested codeine is metabolized to morphine, and many procedures have been described for their simultaneous determination. Various derivatization procedures have been reported including silylation (Medzihradsky & Dahlstrom 1975), pentafluoropropionylation (Combie *et al.* 1982), trifluoroacetylation

(Kintz *et al.* 1987, Hayes *et al.* 1987), heptafluorobutyrylation (Moore 1978) and acetylation (Mathis & Budd 1980). The perfluoroesters are unstable in the presence of moisture owing to the strong negative inductive effects of fluorine atoms in the molecule. Additionally, the trimethylsilyl derivatives are known to be moisture sensitive, and abnormal silylation at the double bond in the allylic alcohol moiety of codeine may take place. Such unusual silylation can also be expected in other opiate compounds having that functionality.

Morphine and codeine are also included in many of the drug screens covered earlier in this chapter. Mule & Casella (1988b) derivatized their extracted sample with pentafluoropentanol and pentafluoropropionic anhydride and heated for 15 minutes at 90°C. The GC procedure is the same as that previously described. Paul *et al.* (1985), having outlined the disadvantages of certain derivatization procedures for some opiates, recommended the method outlined below. They subjected codeine and morphine to acetylation with acetic anhydride; trifluoroacetylation with trifluoroacetic anhydride; pentafluoropropionylation with pentafluoropropionic anhydride; and heptafluorobutyrylation with heptafluorobutyric anhydride. The acetylated derivatives acetylcodeine and diacetylmorphine were found to be the most suitable, being stable at room temperature for 72 hours with an on-column sensitivity of 2 ng. The other derivatives were found to be unsuitable for various reasons. The GC procedure incorporated a fused silica capillary column (15 m × 0.25 mm i.d., DB-5). Operating conditions varied with the compound being analyzed.

Christopherson *et al.* (1987) identified opiates by capillary column GC, using trifluoroacetamide or pentafluoropropionic derivatives. Their procedure successfully determines morphine, codeine, ethylmorphine and 6-monacetylmorphine in urine to a detection limit of 0.25 μmol/l for both derivatives. The chromatographic parameters were as follows: methylsilicone crosslinked capillary column (12.5 m × 0.2 mm i.d.; film thickness 0.33 μm); nitrogen-phosphorus detector operated in the nitrogen mode; column temperature 120 to 220°C at 30°C/minute, then 5°C/minute to 240°C and then to 300°C at 40°C/minute. The samples were injected in the splitless mode and the splitter was reopened after 30 seconds to a split ratio of 1:40.

In a significant number of heroin associated deaths, the concentration of opiate found in the blood of addicts who have apparently died from an overdose, is generally below the minimum concentration (0.2 μg/g blood; 0.2–0.4 μg/g muscle) accepted as fatal (Monforte 1987). For this reason, Pare *et al.* (1984) studied the morphine concentrations in brain tissue from 21 heroin associated deaths. Using a glass column packed with 3% OV-17 on 80.100 mesh Gas-Chrom Q (2 m × 2 mm i.d.), nitrogen as the carrier gas, an isocratic temperature of 235°C and flame ionization detection, they analyzed the bis(*O*-trimethylsilyl) derivative of morphine (Fig. 3.1). They found that in all brain samples the concentration of morphine exceeded the fatal value (0.2 μg/g), but in the corresponding blood samples, only five cases exceeded 0.2 μg/g morphine. The detection limit was 0.07 μg/g of brain tissue. Other workers have since advocated the more routine examination of brain tissue for the presence of drugs of abuse. They suggest that drug concentrations in brain are more meaningful than blood levels, since the brain is the site of action (Spiehler & Reed 1985).

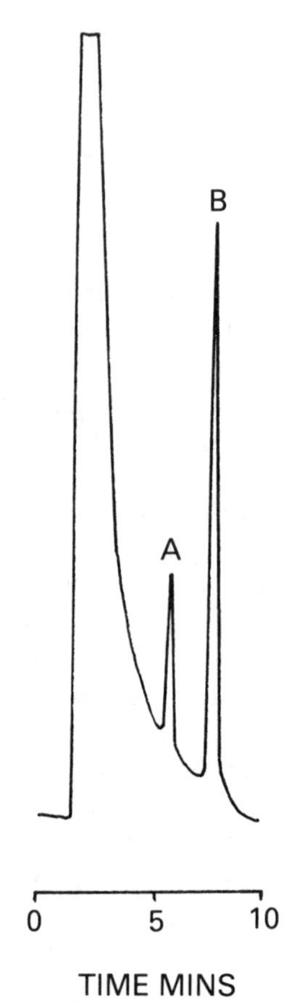

Fig. 3.1. Gas chromatographic separation of the TMS derivatives of morphine (A) and the internal standard nalorphine (B).

3.7 BARBITURATES

The involvement of barbiturates in lethal overdoses has become less significant over the past few years owing to stricter controls and the availability of alternative drugs. The short acting barbiturate, butalbital, is now commonly abused. The determination of phenobarbital is also of importance particularly in cases of sudden death in known epileptics.

GC on packed columns with FID has been widely applied to the simultaneous determination of barbiturates in biological fluids. Chow & Caddy (1986) reported the determination of underivatized barbiturates and other drugs of forensic interest. They prepared three new crosslinkable selective stationary phases by incorporating different monomers, *N,N'*-di(but-3-enyl)amylobarbital, *N,N'*di(pent-4-enyl)amylobarbital and *N,N'*-di(hex-5-enyl)amylobarbital into an SE-54 matrix. The chromatographic performance of these columns was primarily evaluated by using barbiturates, but their versatility was demonstrated by application to benzodiazepines, LSD, narcotic analgesics, tricyclic antidepressants, stimulants and opiates. The separation of 22 underivatized barbiturates commonly encountered in toxicological cases was achieved using the first monomer (Fig. 3.2).

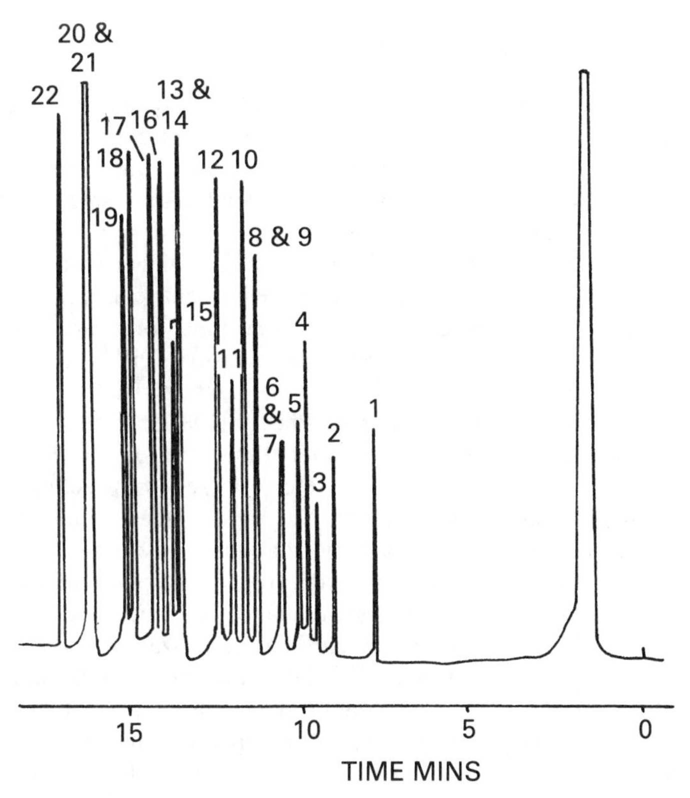

Fig. 3.2. Chromatogram of 22 underivatized barbiturates on a standard SE-54 column.

Suzuki *et al.* (1989) report the determination of nine underivatized barbiturates from human plasma, urine and whole blood, using wide-bore capillary gas chromatography with flame ionization detection. The intermediately polar HP-17 column (10 m × 0.53 mm i.d., film thickness 2.0 μm) was found to give superior chromatographic results to the more commonly used nonpolar phase SPB-1 (15 m × 0.53 mm i.d.). The column temperature was 100–280°C (15°C/minute) and injection temperature 280°C. The limit of detection for this procedure was reported to be 0.5–1.0 μg/ml.

Alternatively, derivatization of the acidic hydrogens in the barbiturate molecules may be performed, in order to reduce their polarity and, hence, their adsorption on the column and peak tailing. Mule & Casella (1989) developed a GC-MS method for the determination of the *N,N′*-dimethyl derivatives of the 5,5′-disubstituted barbiturates. They employed a splitless injection mode and a fused silica capillary column (12.5 m × 0.2 mm i.d.) crosslinked with methylsilicone (0.33 μm film thickness). The GC oven was programmed from 110 to 140°C at 25°C/minute, then to 250°C at 50°C/minute. The detection limit of this system was 20 ng/ml in human urine.

Many different derivatives have been used, including TMS, N-pentafluorobenzyl (PFB). On-column derivitization using trimethylanilinium hydroxide has also been applied to barbiturate analysis.

3.8 BENZODIAZEPINES

The 1,4 and 1,5 benzodiazepines are used as tranquilizers, hypnotics, anticonvulsants or muscle relaxants and are still among the most frequently prescribed drugs even though some of them have disturbing side-effects. They are frequently and increasingly encountered in forensic cases, either alone or in combination with other drugs or alcohol (Jones *et al.* 1985, Parker *et al.* 1988).

Concentrations of benzodiazepines found in urine are higher than those found in plasma, and so urine is usually the matrix of choice for benzodiazepine identification. In urine, metabolites are often excreted as glucuronide conjugates which can be cleaved either by acid or enzyme hydrolysis. Acid hydrolysis yields the corresponding benzophenone, which has been used to identify the parent benzodiazepine using GC-MS. Such a procedure used by Maurer & Pfleger (1987) has been applied to 29 benzodiazepines in human urine 1–2 hours after administration. Their GC procedure used a capillary column (12 m × 0.2 mm i.d.) crosslinked with methylsilicone (film thickness 0.33 μm). The column temperature was programmed from 100 to 310°C at 30°C/minute and the carrier gas was helium. In addition to mass spectral detection, they also employed flame ionization (FID) and nitrogen-phosphorus detection (NPD). An additional advantage of acid hyrolysis is that it eliminates the need to further derivatize the benzodiazepine for determination. (Derivatization is necessary for the thermally labile benzodiazepines). However, some of the newer benzodiazepines such as alprazolam and triazolam do not contain a lactam ring and therefore do not undergo acid-mediated ring conversion to the benzophenone. Jones *et al.* (1988) overcome this problem by modifying the common acid hydrolysis procedure and identify such newer benzodiazepines by using additional mass ions characteristic of the parent drugs in GC-MS analysis. Their capillary GC procedure incorporated a fused silica column (12 m × 0.2 mm i.d.) crosslinked with methylsilicone (film thickness 0.2 μm). The initial temperature remained at 140°C for 2 minutes, then increased to 285°C at 20°C/minute. Helium was used as the carrier gas and the injection mode was splitless. The benzophenones were successfully determined at 0.2 mg/l (the cut-off point for EMIT urine screening assay for benzodiazepines). The authors point out that in most cases, identification of two or more benzophenones characteristic of a particular benzodiazepine or its metabolite provides definitive evidence of the presence of that benzodiazepine. In some cases, however, the benzophenones identified simply limit the number of possible benzodiazepines or metabolites present because some benzodiazepines metabolize to the same intermediates (diazepam and chlordiazepoxide both yield desmethyldiazepam and oxazepam as metabolites) and some yield common benzophenones after acid hydrolysis.

In an attempt to eliminate this problem, Mule & Casella (1989) report the GC-MS determination of trimethylsilyl derivatives of diazolo and triazolo benzodiazepines in human urine. The limit of detection was 50 ng/ml for all diazolobenzodiazepines.

They described two GC systems, one incorporating a capillary column (12 m × 0.2 mm i.d.) crosslinked with dimethylsilicone (0.33 μm film thickness), and the other a 25 m capillary column crosslinked with 5% phenylmethylsilicone. Helium was the carrier gas in both cases, and the injection system was operated in the splitless mode. The sample extracts were derivatized with BSTFA in 1% TMCS.

Sensitivity of detection is an increasing problem when the newer, more potent benzodiazepines are involved. Flunitrazepam, alprazolam and triazolam (single therapeutic doses in the range 0.125–2 mg) in particular have a high affinity for the benzodiazepine receptor as do midazolam and loprazolam. Therefore they require only small amounts to achieve therapeutic results. Because of this, electron capture detectors (ECD) have become popular owing to their increased sensitivity over NPD and FID. Senó *et al.* (1991) used FID and ECD to determine benzophenones after chromatography on an intermediately polar (HP-17) fused silica wide-bore capillary column heated from 180 to 270°C at 5°C/minute. They reported that ECD was 10 times more sensitive than FID. Garzone & Kroboth (1989) described pharmacokinetic parameters for the newer benzodiazepines including alprazolam, triazolam, midazolam and loprazolam. These authors also reported a sensitive and highly specific gas chromatographic method for the determination of these compounds, using an electron capture detector. Other procedures have been described for the separation of these benzodiazepines by using various derivatization procedures (Joern & Joern 1987, Frazer *et al.* 1991). The paper by Joern & Joern reported a limit of detection of 11 ng/ml for alprazolam, 10 ng/ml for its 1-hydroxy metabolite and 12 ng/ml for its 3-hydroxymethyl-5-methyltriazolyl chlorobenzophenone metabolite (Fig. 3.3). Koves & Wells (1989) described the determination of lorazepam in whole blood. In the early literature, lorazepam was determined by using ECD either with no derivatization (Greenblatt *et al.* 1978) or after hydrolysis to the benzophenone. Koves & Wells describe capillary gas chromatography and mass spectrometry with a detection limit of 0.5 ng/ml in whole blood.

3.9 CANNABIS

Although cannabis in sufficiently high doses has some hallucinogenic activity, for the purposes of this discussion it is included along with the sedatives. Cannabis or marijuana is the most widely abused drug (apart from alcohol) in the World. The most popular form of cannabis abuse is smoking. When it is smoked, a variety of physiological and psychological changes are produced.

Cimbura *et al.* (1990) recently carried out a study of the incidence of cannabis and ethanol involvement in fatal road accidents in Ontario, Canada. Using GC to determine cannabis levels, they detected Δ^9-tetrahydrocannabinol (THC) (the main active constituent of cannabis) in the blood of 127 driver victims (10.9%) in concentrations ranging from 0.2 to 37 ng/ml.

The cannabinoids are a large group of compounds unique to cannabis and its extracts, which include tetrahydrocannabinol (THC). This is the compound responsible for the majority of psychotomimetic actions, but interest in the analysis of cannabinoids in biological fluids has also centered on cannabinol (CBN) and

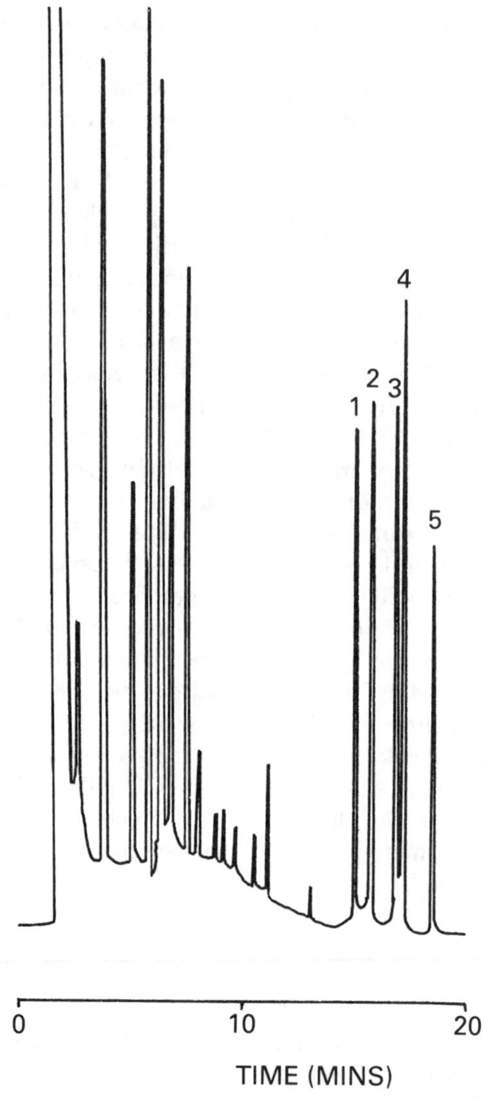

Fig. 3.3. Chromatogram of a urine extract containing approximately 160 ng/ml of (1) 3-hydroxymethyl-5-methyltriazolyl chlorobenzophenone (3-HM), (2) alprazolam, (3) triazolam, (4) an endogenous urinary impurity and (5) 1-hydroxy alprazolam.

cannabidiol (CBD) (Jones *et al.* 1981). THC itself is not normally found in urine, so it is the blood samples which are routinely analyzed because of their greater relevance to brain levels.

The simultaneous determination of THC and its major metabolites 11-nor Δ^9-tetrahydro cannabinol-9-carboxylic acid (THC-COOH) and 11-hydroxy-Δ^9-tetrahydrocannabinol (11-OH THC) in urine and plasma or their individual identification has been the subject of a great deal of research. The blood levels of THC which have to be determined are low (1 ng/ml of blood is a typical value 4–6 hours after smoking). McBurney *et al.* (1986) report levels of THC in plasma to be in excess of 2 to 3 ng/ml, and the metabolite 11-OH THC levels in urine to be in excess of 15 to 20 ng/ml, 6 hours after smoking. At these levels, conventional gas chromatographic systems are no longer sufficiently sensitive, although Baker *et al.* (1984) describe the extraction of THC-COOH from urine and its determination as the trimethylsilyl derivative by using a glass column packed with 3% SP-2250 on 100/200 mesh Supelcoport. However, the majority of recent procedures use open-tubular or capillary columns for cannabinoid analysis. Attempts to improve on packed column sensitivity have generally focused on efficient extraction, silanized glassware (since THC can easily be lost owing to adsorption onto active glass surfaces) and in some cases, the selection of internal standard. If the internal standard is an unnatural homologue of the cannabinoid analyzed for, and is used at a relatively high concentration, the majority of cannabinoids occupying available binding sites will be internal standard (if the concentration of THC is low) and the overall proportion of THC lost to binding will be significantly reduced. Various homologues (such as deuterated analogues, *n*-heptyl Δ^6THC, hexahydrocannabinol) have been used. Several GC methods are available for the determination of THC and/or its metabolites (Whiting & Manders 1982, Hanson *et al.* 1983, McCurdy *et al.* 1986, Paul *et al.* 1987). Above 80°C, the acidic constituents of cannabis can decarboxylate easily unless they are protected as methyl esters or by silylation, and so for this reason most procedures involve some type of derivatization. The screening procedure used by Mule & Casella (1988a) includes the analysis of derivatized THC-COOH. Tetramethyl ammonium hydroxide (TMAH)/dimethylsulfoxide (DMSO) (1:1, v/v) + iodopropane are added to the extract, heated for 15 minutes at 60°C, then cooled. The derivative is then extracted with hexane. An alternative derivatization procedure is to add *n*-methyl-*n*-trimethylsilyl trifluoroacetamide (MSTFA), heat for 15 minutes at 60°C, then inject directly onto the chromatographic column. The GC conditions were as follows: fused silica capillary column (12.5 m × 0.2 mm i.d.) was operated in the splitless mode. The column temperature was 165 to 250°C at 50°C/minute.

Wimbish & Johnson (1990) describe a GC/MS procedure for the determination of Δ^9-carboxy tetrahydrocannabinol, reporting a detection limit of 5 ng/ml and a limit of quantitation of 2.5 ng/ml. They also used TMAH and DMSO to produce the dimethyl derivative prior to determination.

Although the TMS derivative is the most widely used for cannabinoid analysis, ElSohly *et al.* (1984) report a derivatizing procedure for THC-COOH in urine which uses pentafluorobenzyl bromide in a biphasic system, using benzyl tributylammonium hydroxide as a phase transfer catalyst. Detection was by electron capture (Fig. 3.4). Concentrations of THC in the region of 1-2 ng/ml were easily detected by this method.

Foltz & Sunshine (1990) produces the hexafluoroisopropyl/penta fluoropropionyl derivatives of cannabinoids for their detection in urine. They use a chemical ionization procedure rather than the more common electron impact ionization for mass spectrometric detection. Ritchie *et al.* (1987) emphasized the pitfalls of electron capture detection for THC and the expense of mass spectrometers. They recommend the

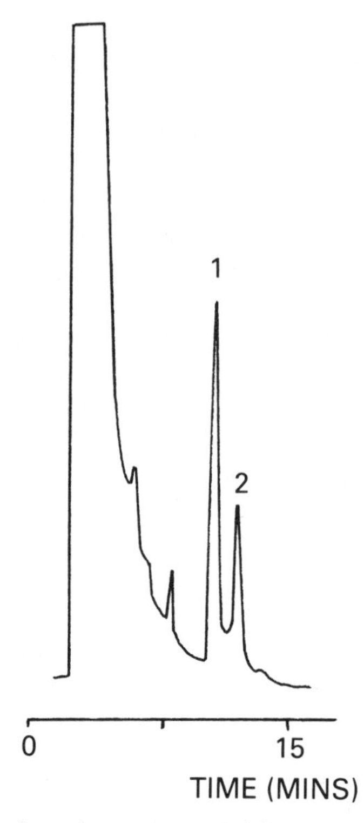

Fig. 3.4. Chromatogram of a urine extract containing (1) Tetrahydro-cannabinolic acid (THC-COOH) and (2) 11-nor-cannabinol-9-carboxylic acid (CBN-COOH) as internal standard. Both derivatized with pentafluorobenzyl bromide.

derivatization of THC with cold 3-pyridine diazonium chloride solution and detection with a nitrogen sensitive detector. Their assay incorporated a capillary (12 m × 0.2 mm i.d., film thickness 0.1 μm OV-101) column, with helium as the carrier and make-up gas. The temperature program ran from 200°C for one minute to 235°C at 20°C/minute, held there for one minute, then increased to 265°C at 5°C/minute. The detection limit was 2 ng/ml of blood.

3.10 ANTIPSYCHOTICS

Of these the largest group consists of the phenothiazine derivatives. Most of these are analogs of chloropromazine. Since most of the phenothiazines are substantially metabolized, absorbed and localized in various tissues, the analysis of these derivatives is necessary in order to fully interpret the effect of the drugs. Negligible amounts of the unchanged drug are excreted in the urine. Chloropromazine can be metabolized to its sulfoxide in addition to demethylated species. Up to 70% of the total dosage of chlorpromazine and phenothiazines is neither recovered nor accounted for.

The tricyclic antidepressants account for the majority of lethal suicide drug ingestions occurring in the United States today and are among the most widely prescribed drugs for the treatment of mental disorders and the therapy of depression. Tricyclics and the newer related tetracyclic drugs are characterized by high lipid solubility, low therapeutic index and poor clinical correlation between administered dosage and resulting drug levels. In the post mortem state, tricyclic antidepressants are released from lipid stores, resulting in erroneously elevated blood levels representing 2–10 times the actual ante-mortem concentration (Apple & Bandt 1988). Even in therapeutic cases, there is a wide variation in steady state blood concentrations between patients receiving the same dose, therefore reproducible, accurate analyses of serum samples is required.

The most frequently used analysis procedures reported in recent years have used GC with NPD, FID or ECD with nitrogen-phosphorus detection being the method most widely used. Van Brunt (1983) reports a method using a fused silica GC capillary column with NPD, as do Abernathy *et al.* (1981). Vinet (1983) reports the determination of underivatized antidepressants by using a nitrogen specific detector. A review of analysis procedures for antidepressants including GC with nitrogen-phosphorus detection was reported by Orsulak *et al.* (1989). A comprehensive GC with nitrogen selective detector screening procedure for nine antidepressants and their metabolites in serum was reported by Rifai *et al.* (1988). They use a wide-bore (60 m × 0.75 mm i.d.) capillary column with a permanently bonded nonpolar stationary phase (SPB-1, 100% polymethylsiloxane, film thickness 1.0 μm). The injector and detector temperatures were set at 325°C and the oven was operated isothermally at 260°C. Helium was the carrier gas with air and hydrogen as detector gases. The limit of

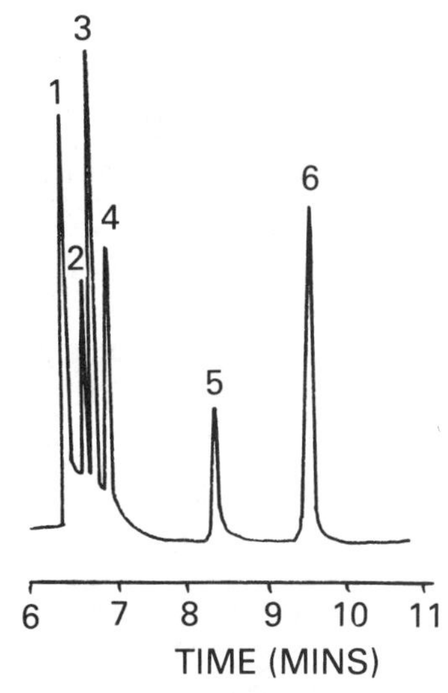

Fig. 3.5. Chromatogram of extracted antidepressant standards: (1) amitriptyline, (2) nortriptyline, (3) imipramine, (4) desimpramine, (5) maprotiline, (6) chlorimipramine.

detection was at least 24 ng/ml for all the drugs investigated (Fig. 3.5). Hattori *et al.* (1990) report the determination of four tricyclic antidepressants using GC with a new type of detector, the surface ionization detector. Their method also uses a SPB-1 capillary column (30 m × 0.32 mm i.d., film thickness 0.25 μm) and a split-splitless injector. The injection temperature was 200°C and the column temperature was run from 100 to 280°C at 6°C/minute. Helium was used as the carrier gas and the sample was injected in the splitless mode. The splitter was opened after two minutes. The detection limit for each drug was 0.5–1.0 ng/ml of sample (5–10 pg per injection volume). This is 2000 times more sensitive than flame ionization detection using a packed column. The surface ionization detector is particularly specific and sensitive

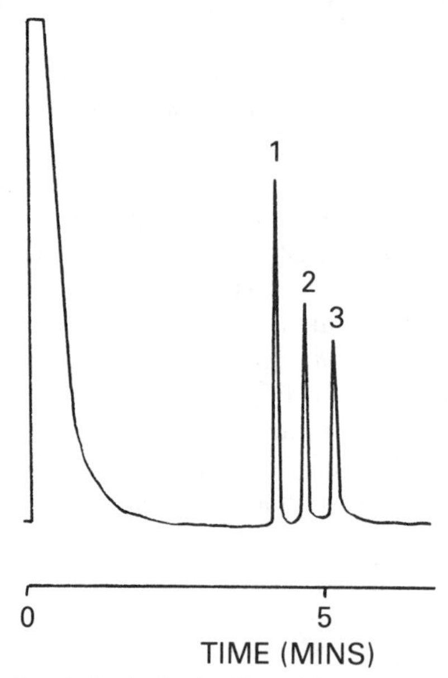

Fig. 3.6. Separation of underivatized tricyclic antidepressants on a phenylpolysiloxane column: (1) amitriptyline, (2) nortriptyline, (3) protriptyline.

for compounds with secondary or tertiary amino groups in their structure. Fig. 3.6 shows a chromatogram of underivatized tricyclic antidepressants on a 15 m × 0.53 mm column with a 1 μm film of 50% phenyl polysiloxane. Direct injection of 1 μl containing 300 ng of each drug in methanol. Helium carrier at 20 ml/minute; 150–220°C at 20°C/minute; FID.

3.11 STIMULANTS

The most significant drugs in this group are cocaine and amphetamine. An estimated 20–30 million Americans have used cocaine at least once and approximately 5–10 million are habitual users. The U.S.A. accounts for 5% of the World's population and 80% of the World's cocaine use. Cocaine probably acts by activation of the brain pleasure centres which are dependent on dopamine neurotransmission. Cocaine blocks the re-uptake of dopamine from the synapse which results in euphoria, increased motor activity and psychotic symptoms. The attraction of the drug is obviously its stimulant effects; however, side-effects include increased heart rate, cardiac arrhythmias and peripheral vasoconstriction. Consequently, sudden changes in blood pressure may result in death due to cardiac failure and cerebral hemorrhage.

Amphetamine is similar in pharmacological effects and is also thought to act via its interaction with dopamine. Amphetamine promotes the release of dopamine and norepinephrine from presynaptic neurones and blocks the uptake of catecholamines. Like cocaine, amphetamine causes restlessness, stimulation, appetite suppression, paranoia and psychosis. Other stimulants with amphetamine-like activity include methamphetamine, phenmetrazine, dexamyl, drinamyl, mephentermine, pipradol and methylphenidate.

Again several commonly prescribed and over the counter drugs are abused because of their stimulant activity. Cough/cold remedies and diet preparations frequently contain stimulant drugs. Ephedrine stimulates both alpha and beta adrenergic receptors and has some CNS stimulant activity. Phenylephrine is similar in action, and phenylpropanolamine, which is a common ingredient of diet pills, is a potent stimulant. Diethylpropion, mazindol, phentermine, benzphetamine and phendimetrazine are other stimulants frequently encountered.

3.12 COCAINE

Cocaine is rapidly metabolized in the body. 14–17% of cocaine dose is excreted in the urine as benzoylecgonine (BZE), and 12–21% as ecgonine methyl ester (EME) (Ambre *et al.* 1988). Other reported metabolites include methylecgonine, norcocaine, ecgnonine, benzoylnorecgonine, ecgonidine methyl ester and hydroxy-methoxy substituted BZE. Recently, four new metabolites have been reported (Zhang & Foltz 1990). Further, cocaine has been shown to be unstable in blood or in aqueous solutions, and so, from a forensic point of view, it may be difficult to draw conclusions from the concentrations found in post mortem blood regarding the concentration of cocaine in the body at the time of death. Hence, the majority of analysis methods are designed to detect cocaine in addition to its metabolites in body fluids.

Gas chromatographic procedures for the determination of cocaine and its metabolites in biological fluids are widespread. Joern (1987), Taylor *et al.* (1987) and Jacob *et al.* (1987) all report GC procedures with detection limits of 35 ng/ml, 50 ng/ml and 10 ng/ml respectively for cocaine and BZE in urine or plasma.

Cocaine and EME can be determined by CG without the need for derivatization by using mass spectrometric (Mule & Casella 1988a), electron capture, or nitrogen-phosphorus detection (McCurdy 1980), but its metabolites, particularly BZE and ecgonine, show greatly improved chromatographic characteristics when derivatized prior to determination. Many derivatization procedurers have been reported. Jacob *et al.* (1987) derivatized BZE to the butyl ester. Taylor *et al.* (1987) converted BZE to its trimethylsilyl derivative, using MSTFA. Spiehler & Reed (1985), when analyzing brain samples, converted BZE to the propyl ester derivative, using dimethyl formamide dipropyl acetal in dimethyl formamide. Meeker & Reynolds (1990) also used this derivatization procedure for their analysis of BZE in fetal blood. Zhang & Foltz (1990) derivatized urine extracts with pentafluoro propionic anhydride (PFPA) and 1,1,1,3,3,3-hexafluoro isopropanol (HFIP), heated for 10 minutes at 70°C and evaporated prior to analysis. This derivatization procedure produced the pentafluoropropionyl derivative of ecgonine methyl ester; the hexafluoroisopropyl derivative of BZE; the N-pentafluoropropionyl derivative of norcocaine; hexafluoropropyl pentafluoropropionyl derivative of ecgonine. Cocaine and ecgonidine methyl ester were unaffected by the derivatization procedure. The GC parameters for the analysis were as follows: DB-1 capillary column (15 m × 0.25 mm i.d.) with a film thickness of 0.25 μm. Hydrogen was used as the carrier gas and the oven temperature was 100°C. After a splitless injection, the temperature was maintained for 0.5 minutes, then with the split flow turned off, programmed to 250°C at 10°C/minute.

Verebey & DePace (1989) determined BZE in urine samples by GC-NPD by first derivatizing it with pentafluoropropionic anhydride and pentafluoropropanol at 90°C for one minute. A DB-1 fused silica capillary column (15 m × 2 mm i.d.; film thickness 0.25 μm) was used for the chromatography and the detection limit was 300 ng/ml of BZE. The majority of cocaine determination procedures are run on capillary columns (12 or 15 m × 0.2 mm i.d.), coated with DB-1 or DB-5 methylsilicones, using splitless injection and temperature programs between 150 and 280°C. Helium or hydrogen is used as the carrier gas. Jacob *et al.* (1987) used both capillary and packed column GC to determine BZE, and capillary GC only for cocaine analysis. His capillary column was a narrow-bore (12 m × 0.2 mm i.d.) fused silica column crosslinked with methylsilicone (0.33 μm film thickness). The oven temperature was programmed from 120 to 250°C at 32°C/minute and held at the high temperature for 4 minutes. Injections were made in the splitless mode. The conventional system consisted of a 6 m glass column packed with 3% SP2100 DB on Supelcoport (100–120 mesh), and the oven was run isothermally at 250°C. The carrier gas (helium) flowrates were 1 ml/minute for the capillary column analyses and 30 ml/minute for packed column analyses. Using *m*-toluylecgonine as an internal standard, both cocaine and the derivative could then be determined on either capillary or packed columns (Fig. 3.7).

Recently, a new and highly significant metabolite has been detected by GC-MS in the urine of cocaine users in the presence of ethanol. It has been identified as cocaethylene and is formed only in the presence of ethanol. Its concentration in blood frequently approaches, and occasionally surpasses, that of cocaine (1.6 mg/l). Additionally, it has been shown to be formed in the liver from cocaine and ethanol, and is pharmacologically active (Rose *et al.* 1991). It is determined by the same GC procedures as those used for cocaine.

3.13 AMPHETAMINE AND METHAMPHETAMINE

Amphetamine and methamphetamine are powerful CNS stimulants which produce euphoric effects. Procedures for the determination of therapeutic levels of underivatized amphetamine have been reported by Christopherson *et al.* (1988) (whole blood), by Taylor *et al.* (1989) (urine) and by Kintz *et al.* (1989) (plasma and urine). Christopherson used two systems, one incorporating a packed column (GP 10% Apiezon L/2% potassium hydroxide on 80-100 mesh Chromosorb W AW; 180 cm × 2 mm i.d.) and one using a capillary column (crosslinked methylsilicone; 12.5 m × 0.2 mm i.d., 0.32 μm film thickness). For the packed system, The injector

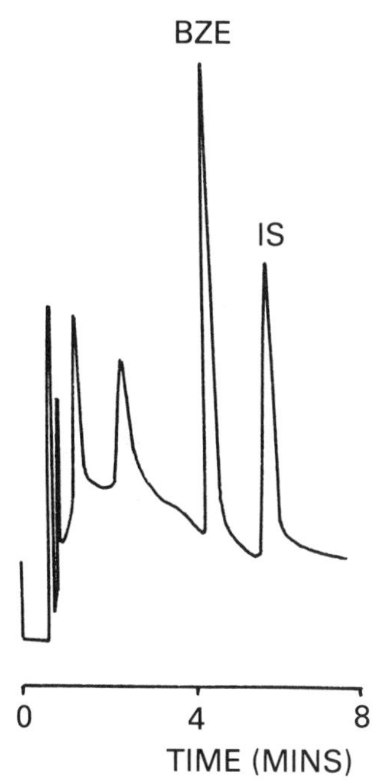

Fig. 3.7. Chromatogram of an extract of human plasma containing benzoylecgonine (BZE) and *m*-toluylecgonine as internal standard (IS).

temperature was set at 230°C, detector 245°C and the oven was operated at an isothermal temperature of 170°C. Nitrogen was used as the carrier gas, and hydrogen and air were used as detector gases for the nitrogen-phosphorus detector operated in the nitrogen mode.

For the capillary system, the injector was operated in splitless mode but the splitter was reopened after 30 seconds. The injector and detector temperatures were both 250°C and the column was operated from 60°C (30s) to 130°C at 10°C/minute, and then 40°C/minute to 250°C. Helium was used both as the carrier gas and the make-up gas. Hydrogen and air were used as detector gases. Satisfactory chromatograms were obtained by using both systems without the need for derivatization, and only 0.5 ml of blood was required for determination (Fig. 3.8). The detection limits for the packed column and capillary systems were 0.2 and 0.1 μmol/l respectively. The packed system is used routinely since it gives the best precision and most stable chromatographic conditions, but the capillary system is used for samples containing other drugs or metabolites.

Taylor used a CB-5 methylphenyl polysiloxane fused silica capillary column (15 m × 0.25 mm; 0.25 μm film thickness), a split injector (ratio 20:1), an isothermal oven set at 150°C and injector and detector temperatures of 250 and 280°C respectively. Drugs in their free form were determined by using a nitrogen-phosphorus detector, but for mass spectrometry, the heptafluorobutyric anhydride derivatives were formed. The detection limit for amphetamine and methamphetamine with this

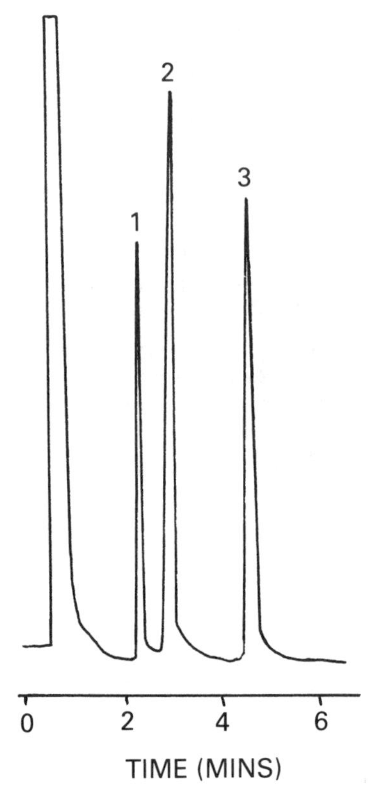

Fig. 3.8. Chromatogram of a whole blood extract spike with (1) amphetamine, (2) methylamphetamine and quinoline (3) as internal standard.

procedure was 35 ng/ml. Kintz determined CNS stimulants, including amphetamine, in plasma and urine, using a capillary column to a detection limit of 4 ng/ml.

Ante-mortem blood and urine were analyzed for methamphetamine by Rasmussen *et al.* (1989) using a 2.5 mm × 20 m capillary column, an isothermal temperature program (175°C) and helium as the carrier gas. An isothermal procedure for the determination of methamphetamine and its metabolites in monkey urine was reported by Yamamoto *et al.* (1989). They used a glass capillary column (2 mm × 2 m), packed with 2% OV-17 on a Uniport HP 80/100 mesh. The oven temperature was 150°C. Amphetamine and methamphetamine are also covered in the analysis of stimulants by GC using nitrogen-phosphorus and mass spectrometric detection procedures undertaken at the 1988 Olympic Games in Seoul, Korea (Lho *et al.* 1990).

Amphetamine, methamphetamine and related compound analysis usually requires derivatization of the drugs particularly when mass spectrometry is required. The most common derivatizing agents are trifluoroacetic anhydrides (TFA) (Mule & Casella 1988a), *N*-trifluoroacetyl-L-prolyl chlorides (Fitzgerald *et al.* 1988), heptafluoro derivatives (Lillsunde & Korte 1991) or trichloroacetic anhydrides (Hornbeck & Czarny 1989). Hornbeck & Czarny tested 11 different derivatizing agents and procedures before concluding that trichloroacetyl derivatives are the best for the analysis of amphetamines.

3.14 HALLUCINOGENS

3.14.1 Lysergic acid diethylamide (LSD)

LSD is a very potent psychoactive drug that is commonly abused. The level of LSD normally consumed to effect hallucinations is between 40 and 120 μg. It is rapidly metabolized, with less than 1% of unchanged LSD being excreted in the urine. Because of this, and its instability in acid, heat and light, it is a very difficult drug to identify in biological fluids. Formation of the trimethylsilyl derivative improves the volatility and gas chromatographic characteristics of LSD. However, the TMS derivative degrades rapidly in the presence of water or aqueous solvents. When protected from moisture, it is stable for several days at temperatures between 25 and −25°C.

Francom *et al.* (1988) report a single step extraction followed by GC-MS determination. They formed the *N*-trimethylsilyl derivative of LSD using BSTFA. Extractions and derivatization procedures were carried out with fluorescent room lighting because of the sensitivity of LSD to daylight. GC was performed on a crosslinked dimethyl silicone fused silica capillary column (12.5 m × 0.2 mm i.d.; 0.33 μm film thickness). Splitless injection with a split valve off time of 0.9 minutes was employed. The oven was programmed from 200 to 320°C at a rate of 20°C/minute and maintained at the final temperature for 6 minutes. Absorptive loss of LSD on chromatographic columns is a problem. Derivatization and the use of capillary columns help in addressing this phenomenon, but generally, column conditioning or deactivation must be carried out before analysis. Francom *et al.*, deactivated the column by using five successive injections of BSTFA. This resulted in a threefold increase in detector response. Silanization of the glass injection port liners was also important, increasing sensitivity by 20% when carried out. Their reported detection limit was 0.5 ng/ml of urine.

Paul *et al.* (1990) have reported repeatedly observing co-elution of interferences when using Francom's procedure. Paul *et al.* describe GC-MS analysis using a capillary column (15 m × 0.25 mm i.d.; DB5), helium carrier gas flowing at 1.26 ml/minute and an initital temperature of 190°C rising to 290°C at 20°C/minute. The final temperature was maintained for 3 minutes. Injection was performed in the splitless mode, but after 30 seconds, the instrument was turned to split mode (1:30) to purge any solvent that could cause tailing. Prior to analysis, the extracts were converted to trimethylsilyl derivatives with BSTFA. The derivative was stable for at least seven days when stored at 0–5°C. The authors point out that LSD is sensitive to silicic acid and therefore any amount of this present on the GC column will adversely affect the sensitivity of detection. They recommend conditioning of the column prior to sample injection by injecting 2–3 μl of derivatized urine extract four times. The basic components of the urine conditioned the column satisfactorily. Their procedure was used to determine LSD in urine, and they report a detection limit of 29 pg/ml although considerable interferences were observed when the LSD concentration was less than 100 pg/ml. Papac *et al.* (1990) have measured LSD in plasma by using GC, negative ion chemical ionization-mass spectrometry. They used trifluoroacetylimidazole to produce the N-trifluoroacetyl derivative. A general purpose capillary column was used for the analysis and the oven was programmed from 180 to 300°C at 20°C/minute. The reagent gas was methane.

3.14.2 Phencylidine (PCP)

Phencyclidine (phenylcyclohexyl piperidine or PCP) was originally marketed as a veterinary anaesthetic. Now used illicitly as a white powder which is taken orally, smoked often together with cannabis, or snorted in combination with cocaine. Symptoms appear rapidly and resemble schizophrenia, with users becoming a risk to themselves and to others. Euphoria, depression, agitation, violence, hallucinations, paranoia, panic and suicidal tendencies are effects commonly caused by this drug. The PCP user also experiences increased strength and a decreased sense of pain. This makes the person extremely dangerous and difficult to control, before eventually losing consciousness and going into coma. Hypertension and tachycardia are also associated with the use of PCP, which may result in cardiac failure. PCP induced psychosis can last for up to 4 weeks, during which time the addict has to be kept sedated because of violent behaviour. Phencyclidine is thought to act by increasing glucose utilization in the brain. This is probably initiated via a receptor and may involve acetylcholine. PCP intoxication is characterized by reddening of the skin, enlarged pupils, delusions, amnesia, nystagmus, excitement, arrhythmias, paranoid psychosis and violent behaviour. With high doses, convulsions develop which lead to death. Phencylidine is a commonly abused drug in the U.S.A. (Bailey 1987, Ahmad 1987, Garey *et al*, 1987). It is rapidly metabolized in the body and excreted in urine mainly as metabolites. In forensic cases, however, the parent drug is often detected. Several GC methods have been developed for the determination of phencyclidine and its metabolites in biological fluids and it is included in several drug screening procedures where derivatization is not required (Mule & Cassella (1988), Kintz *et al.* (1990)).

Verebey & DePace (1988) describe a rapid method for the confirmation of PCP in urine following EMIT assay. They use a capillary column (15 m × 0.2 mm i.d. WCOT fused silica with 0.25 μm DB1 coating) and a flame ionization detector. The column temperature was programmed from 160 to 190°C at a rate of 30°C/minute and the injection and detector ports were held at 300°C. Their detection limit was reported as 5–10 ng of extracted PCP.

The development of sufficiently sensitive procedures for the determination of phencyclidine has received much attention. Woodworth *et al.* (1984) report a sensitive procedure for the determination of PCP, its monohydroxy metabolites and a pentanoic acid metabolite. The *cis* and *trans* isomers of the 1-(1-phenyl-4-cyclohexyl) piperidine (PPC) metabolite were derivatized with BSTFA and TCMS, then separated by capillary column GC. A slightly different extraction procedure and GC method was required for the determination of the pentanoic acid metabolite.

The detection of low nanogram levels of phencyclidine (PCP) extracted from biological fluids, particularly urine, has generally been difficult to achieve by conventional gas chromatography. This is because the adsorptive properties of most GC column packing support systems decrease the sensitivity of the systems towards alkaloids. Street *et al.* (1977) successfully solved this problem by heat treating the acetylated support and incorporating nitrogen-phosphorus detection. (The nitrogen-phosphorus detector has since been shown to be ten to fifty times more sensitive to alkaloids than flame ionization detectors).

Kandiko *et al.* (1990) modified the procedure of Street and applied it to the determination of phencyclidine in urine. Their detection limit was 15 ng/ml. The glass column (91.4 cm × 2 mm i.d.) was packed with Chromosorb W AW-DMCSW (40 g), washed in 0.35 mol/l potassium hydroxide and allowed to stand for 1 hour. The mixture was boiled for 10 minutes and rinsed with distilled water. The water was decanted from the support material and the support was air-dried. A pyridine-acetic anhydride mixture (3:2 v/v) (100 ml) was added and allowed to stand at room temperature for 48 hours. The acetylated support was then rinsed with acetone and air-dried. Then 10% SE-30 in benzene (3:2 v/v) was added and the support was allowed to stand at room temperature for 48 hours. Post drying, the coated support was transferred to a 20 cm × 4 mm i.d. glass column. The glass tube was wrapped in aluminium foil and heated for 1 hour under nitrogen. The glass column was treated in a similar manner to the support material. Although a lengthy procedure, once completed, the heat treated acetylated column was sensitive to PCP for one year at column temperatures of 230°C.

The increase in column sensitivity caused by the treatment described is probably due to reduced hydrogen bonding between the support and PCP. Street indicated that acylation of diatomaceous earth eliminated the first of the two types of hydrogen bonding inherent in the support, the Si-OH terminal hydroxy bonding which acts as a proton donor for hydrogen bonding. The second, Si-O-Si, acts as a proton acceptor. This interaction was deemed impossible by Kandiko owing to the high molecular weight of compounds such as PCP. (Street has previously shown that high molecular weight alkaloids such as morphine do not form hydrogen bonds with the Si–O–Si moiety owing to stearic hindrance.)

Phencyclidine, its metabolites and derivatives, were determined in plasma, urine and tissue homogenates by using conventional packed column GC with nitrogen-phosphorus detection by Holsztynska & Domino (1986) to a detection level of 5 pmol per injection. Their procedure included derivatization of the water soluble metabolites with heptafluorobutyric anhydride prior to analysis. The column (1.5 m × 2 mm i.d.; 3.5% SE-30 on GasChrom Q,100/200 mesh) was operated isothermally at 180°C (Fig. 3.9).

3.15 MISCELLANEOUS

In addition to the common drugs of abuse which have been outlined above, both anti-inflammatory agents and, increasingly, steroids are frequently encountered in the forensic toxicology laboratory.

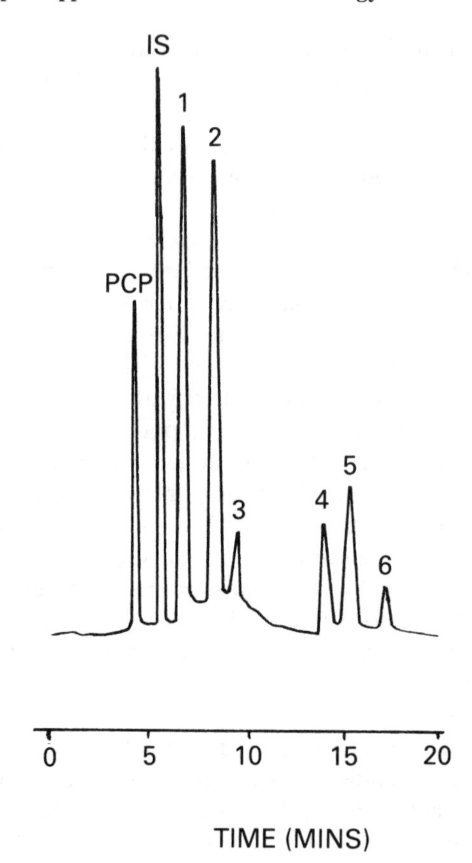

Fig. 3.9. Gas chromatographic separation of PCP and heptafluorobutyric anhydride derivatives of mono and dihydroxy metabolites from urine. (1) 3-OH cyclo PCP (trans), (2) 4-OH cyclo PCP (*cis* and *trans*), (3) 4-OH PCP, (4) possibly 3,4, diOH cyclo PCP, (5) possibly 3-OH cyclo 4-OH PCP, (6) 4-OH cyclo 4-OH PCP.

3.15.1 Anti-inflammatory drugs

Aspirin and acetaminophen (paracetemol) have been the most commonly abused non-steroidal anti-inflammatory drugs in recent years. However, ibuprofen, ketoprofen and indomethacin are now widely available as prescription drugs and, as such, are liable to abuse. General analysis procedures for anti-inflammatory drugs by GC include determination of their methyl derivatives on packed GC columns (3% SE-30 on 80-100 mesh Chromosorb G, acid washed and dimethyldichlorosilane treated), with nitrogen as the carrier gas and running the temperature program from 120 to 260°C at 20°C/minute. (For acetaminophen analysis, the temperature should be increased to 300°C). Methyl derivatives should be formed by heating the acidic extract with iodomethane and potassium carbonate for 30 minutes at 100°C. Trimethylanilinium hydroxide is not recommended for methylating these compounds.

More recently, the development of a procedure to determine subnanogram quantities of indomethacin in plasma (Nishioka *et al.* 1990) and synovial fluid has been reported (Dawson *et al.* 1990). Dawson forms the pentafluorobenzyl ester of indomethacin prior to determination by capillary GC-MS. The limit of detection was 0.1 ng/ml.

Blesssington *et al.* (1989) form the R-alpha-phenylethylamide diastereomeric derivatives of anti-inflammatory drugs and herbicides before determining them by packed column GC (3% OV-1 on Chromosorb G). They report that analysis times were long, and much faster and better chromatography was achieved by using capillary GC (bonded methylsilicone).

3.15.2 Anabolic steroids

Recently, side-effects and psychomimetic trends have surfaced as a result of anabolic steroid abuse. Shaposhnikov *et al.* (1990) suggested that impaired steroid hormone secretion and metabolism may be the cause of symptomatic hypertension, and Pope & Katz (1990) describe the case of three men with no history of violence or social disorders, who impulsively committed violent crimes, including murder, while taking anabolic steroids. Structured psychiatric interviews suggested that steroids played a significant, if not primary role in the etiology of the violent behaviour.

In 1974, anabolic steroids were added to the list of doping agents banned by the International Olympic Committee (IOC). Nevertheless, steroid use among athletes appears to be increasing rather than decreasing. Hatton & Catlin (1987) reported the GC-MS procedures used for steroid detection as required by the IOC, the USOC and, more recently, the National Collegiate Athletic Association.

Many GC-MS methods exist for the determination of steroids either qualitatively or quantitatively (Vrbanac *et al.* 1982, Axelson *et al.* 1981, Cartoni *et al.* 1983).

Steroids can be excreted in their free form or as conjugated glucuronides. Chung *et al.* (1990) determined both conjugated and free anabolic steroids in urine by using GC-MS. The free steroids were derivatized with MSTFA-TMCS-TMS imidazole (100:5:2,v/v/v) and heated to 80°C for 5 minutes. Then N-methyl bis-heptafluorobutyramide (MBHFB) was added and the mixture heated for 10 minutes. Chromatography was carried out on a crosslinked 5% phenylmethylsilicone capillary column (17 m × 0.2 mm i.d.; film thickness 0.33 μm). The oven was programmed from 180 to 300°C at 25°C/minute. The carrier gas was helium (0.7 ml/minute), and sample extracts were injected in the splitless mode. The conjugated steroids were first enzyme hydrolysed with B-glucuronidase and dithioerthritol. To derivatize the hydroxyl groups only, MSTFA-TMCS (100:2) was added and the mixture heated at 60°C for 3 minutes. Chromatography was carried out on a fused silica capillary column (17 m × 0.2 mm i.d.; film thickness 0.11 μm). The oven was programmed from 180 to 224°C at 4°C/minute, then to 300°C at 15°C/minute. The carrier gas was hydrogen (1.2 ml/minute) and the split ratio was 1:10.

A general chromatographic procedure for the determination of steroids involves the analysis of silyl or methoxime-silyl derivatives. These are produced by heating the sample extract with an 8% solution of methoxyamine hydrochloride in dry pyridine for 30 minutes at 60°C. The evaporated residue is then treated with BSTFA and TCMS to form trimethylsilyl derivatives. A capillary GC column (12.5 m × 0.2 mm i.d.) of fused silica should be programmed from 150 to 200°C at

20°C/minute, then to 260°C at 5°C/minute, using helium as the carrier gas and operating in the splitless mode (Houghton & Teale 1981).

3.16 CONCLUSION

This chapter has attempted to briefly describe the types of drugs encountered in the forensic toxicology laboratory and has reviewed the use of gas chromatography in the examination of these compounds. Both standard 'tried and trusted' procedures and those methods described in the most recent literature have been reviewed. It is apparent that gas chromatography, particularly in combination with the mass spectrometer, will continue to have an important role to play in the foreseeable future for the examination of biological samples for drugs. To this end, a great deal of progress has been made over the last few years in the development and continual refinement of instrumentation and methodologies. The detection of nanogram quantities of drugs is now routine and, as instruments become more sophisticated, offering even greater sensitivity and resolution, so more reports of 'new' metabolites appear. The problem facing the forensic toxicologist is quickly approaching the point where detection of drugs and metabolites is no longer difficult, but what does it all mean? It is the interpretation of the data which is becoming the main concern of the courts, and a strong background in pharmacology and pharmacokinetics in addition to analytical chemistry will soon be required of all forensic toxicologists.

REFERENCES

Abernathy, D.R., Greenblatt, D.J., Shader, R.I. (1981), *Pharmacology* **25** 57–63.

Ahmad, G. (1987), *J. Toxicol. Clin. Toxicol.* **25** (4) 341–346.

Ambre, J. & Ruo, T.I., Nelson, J., & Belknap, S. (1988), *J. Anal. Tox.* **12** (6) 301–306.

Apple, F.S. & Bandt, C.M. (1988), *Am. J. Clin. Pathol.* **89** 794–796.

Ardrey, R.E. & Moffat, A.C. (1981), *J. Chromatogr.* **220** 195–252.

Axelson, M., Sahlberg, B.L. & Sjobvall, J. (1981), *J. Chromatogr.* **224** 355–370.

Bailey, D.N. (1987), *J. Toxicol. Clin. Toxicol.* **25** (6) 517–526.

Baker, T.S., Harry, J.V., Russell, J.W. & Meyers, R.L. (1984), *J. Anal. Tox.* **8** 255–259.

Blessington, B., Crabb, N., Karkee, S. & Northage, A. (1989), *J. Chromatogr.* **469** 183–190.

Bowie, L.J. & Kirkpatrick, P.B. (1989), *J. Anal. Tox.* **13** 326–329.

Caldwell, R. & Challenger, J. (1989), *Ann. Clin. Biochem.* **26** 430–443.

Cartoni, G.P., Ciardi, M., Giarusso, A. & Rosati, F. (1983), *J. Chromatogr.* **279** 515–522.

Chen, X., Wijsbeek, J., Van Ween, J., Franke, J.P. & de Zeeuw, R.A. (1990), *J. Chromatogr.* (*Biomedical Applications*), **529** 161–166.

Chow, N.M.L. & Caddy, B. (1986), *J. Chromatogr.* **354** 219–229.

Christopherson, A.S., Biseth, A., Skuterud, B. & Gadeholt, G. (1987), *J. Chromatogr.* (*Biomedical Applications*), **422** 117–124.

Christopherson, A.S., Dahlin, E. & Pettersen, G. (1988), *J. Chromatogr.* (*Biomedical Applications*), **432** 290–296.

Chung, B.C., Choo, H.-Y.P., Kim, T.W., Eom, K.D., Kwon, O.S., Suh, J., Yang, J. & Park, J. (1990), *J. Anal. Tox.* **14** 91–95.

Cimbura, G., Lucas, D.M., Bennett, R.C., Donelson, A.C. (1990), *J. Forens. Sci.* **35** (5) 1035–1041.

Combie, J., Blake, J.W., Nugent, T.E. & Tobin, T. (1982), *Clin. Chem.* **28** 83–86.

Cox, R.A., Crifasi, J.A., Dickey, R.E., Ketzler, S.C. & Pshak, G.L. (1989), *J. Anal. Tox.* **13** 224.

Dawson, M., Smith, M.D. & McGee, C.M. (1990), *Biomed. Environ. Mass Spectrom.* **19** (8) 453–458.

Eklund, A., Jonsson, J. & Schuberth, J. (1983), *J. Anal. Tox.* **7** 24–28.

ElSohly, M.A., Arafat, E.S. & Jones, A.B. (1984), *J. Anal. Tox.* **8** 7–9.

Fehn, J. & Megges, G. (1985), *J. Anal. Tox.* **9** 134–138.

Fitzgerald, R.L., Blanke, R.V., Glennon, R.A., Yousif, M.Y., Rosecrans, J.A., Francom, P., Andrenyak, D., Lim, H.-K., Bridges, R.R., Foltz, R.L. & Jones, R.T. (1988), *J. Anal. Tox.* **12** 1–5.

Fraser, A.D., Bryan, W. & Isner, A.F. (1991), *J. Anal. Tox.* **15** 8–11.

Foltz, R.L. & Sunshine, I. (1990), *J. Anal. Tox.* **14** (6) 375–378.

Garey, R.E., Daul, G.C., Samuels, M.S., Avery Ragan, F. & Hite, S.A. (1987), *Am. J. Drug Alcohol Abuse*, **13** (1,2) 135–144.

Garzone, P.D. & Kroboth, P.D. (1989), *Clin. Phamacokinet.* **16** (6) 337–364.

Gibb, pers. comm.

Greenblatt, D.J., Franke, K. & Shader, R.I. (1978), *J. Chromatogr.* **146** 311–320.

Hanson, V.W., Buonarati, M.H., Baselt, R.C., Wade, N.A., Yep, C., Biasotti, A.A., Reeve, V.C., Wong, A.S. & Orbanowsky, M.W. (1983), *J. Anal. Tox.* **7** 96–102.

Hatton, C.K. & Catlin, D.H. (1987), *Clin. Lab. Med.* **7** (3) 655–658.

Hayes, L.W., Krasselt, W.G. & Meuggler, P.A. (1987). *Clin. Chem.* **33** (6) 806–808.

Holsztynska, D.J. & Domino, E.G. (1986), *J. Anal. Tox.* **10** 107–115.

Hornbeck, C.L. & Czarny, R.J. (1989), *J. Anal. Tox.* **13** 144–149.

Houghton, E. & Teale, P. (1981), *Biomed. Mass Spectrom.* **8** 358–361

Jacob, P., Elias-Baker, B.A., Jones, R.T. & Benowitz, N.L. (1987), *J. Chromatogr.* **417** (2) 277–86.

Jaeger, H. (1985), *Glass capillary Chromatography in clinical medicine and pharmacology*, Marcel Dekker Inc.

Joern, W.A. (1987), *J. Anal. Tox.* **11** (3) 110–112.

Joern, W.A. & Joern, A.B. (1987), *J. Anal. Tox.* **11** (3) 247–251.

Kandiko, C.T., Browning, S., Cooper, T. & Cox, W. (1990), *J. Chromatogr.* **528** 208–213.

Kintz, P., Mangin, P., Lugnier, A.A. & Chaumont, A.J. (1987), *Eu. J. Clin. Pharmacol.* **37** 531–532.

Kintz, P., Tracqui, A., Mangin, P., Lugnier, A.A. & Chaumont, A.J. (1989), *Forens. Sci. Int.* **40** (2) 153–159.

Kintz, P., Tracqui, A., Lugnier, A.A., Mangin, P. & Chaumont, A.J. (1990), *Methods Find. Exp. Clin. Pharmacol.* **12** (3) 193–196.

Koves, E.M. & Yen, B. (1989), *J. Anal. Tox.* **13** 69–72.
Lho, D.S., Shin, H.S., Kang, B.K. & Park, J. (1990), *J. Anal. Tox.* **14** 73–76.
Lillsunde, P. & Korte, T. (1991), *J. Anal. Tox.* **15** 71–81.
Manca, D., Ferron, L. & Weber, J.-P. (1989), *Clin. Chem.* **35** (4) 601–607.
Mathis, D.F. & Budd, R.D. (1980), *Clin. Toxicol.* **16** 181–188.
McBurney, L.J., Bobbie, B.A. & Sepp, L.A. (1986), *J. Anal. Tox.* **10** 56–63.
McCurdy, H.H. (1980), *J. Anal. Tox.* **4** 82.
McCurdy, H.H., Lewllen, L.J. & Callhan, L.S., Childs, P.S. (1986), *J. Anal. Tox.* **10** 175–177.
Medzihradsky, F. & Dahlstrom, P.J. (1975), *Pharmacol. Res. Commun.* **7** 55–69.
Meeker, J.E. & Reynolds, P.C. (1990), *J. Anal Tox.* **14** 379–382.
Monforte, J.R. (1978), *J. Forens. Sci.* **22** 718–724.
Moore, J.M. (1978), *J. Chromatogr.* **147** 327–336.
Mule, S.J. & Casella, G.A. (1988a), *J. Anal Tox.* **12** 102–107.
Mule, S.J. & Casella, G.A. (1988b), *Clin. Chem.* **34** 1427–1430.
Mule, S.J. & Casella, G.A. (1989), *J. Anal Tox.* **13** (3) 179–184.
Nishioka, R., Harimoto, T., Umeda, I., Yamamoto, S., & Oi, N. (1990), *J. Chromatogr.* **526** (1) 210–214.
Orsulak, P.J., Haven, M.C., Burton, M.E. & Akers, L.C. (1989), *Clin. Chem.* **35** (7) 1318–1325.
Papac, D.I. & Foltz, R.L. (1990), *J. Anal. Tox.* **14** (3) 189–190.
Pare, E.M., Monforte, J.R. & Thibert, R.J. (1984), *J. Anal. Tox.* **8** 213–216.
Park, J., Park, S., Lho, D., Choo, H.P., Chung, B., Yoon, C., Min, H. & Choi, M.J. (1990), *J. Anal. Tox.* **14** 66–72.
Paul, B.D., Mell, L.D., Mitchell, J.M., Irving, J. & Novak, A.J. (1985), *J. Anal. Tox.* **9** 222–225.
Paul, B.D., Mell, L.D., Mitchell, J.M. & McKinley, R.M. (1987), *J. Anal. Tox.* **11** 1–5.
Paul, B.D., Mitchell, J.M., Mell, L.D. & Irving, J. (1989), *J. Anal. Tox.* **13** 2–7.
Paul, B.D., Mitchell, J.M., Burbage, R., Moy, M. & Sroka, R. (1990), *J. Chromatogr.* (*Biomedical Applications*) **529** 103–112.
Poklis, A. (1989), *J. Chromatogr.* (*Biomedical Applications*), **490** 59–69.
Pope, H.G. & Katz, D.L. (1990), *J. Clin. Psychiatry* **51** (1) 28–31.
Rasmussen, S., Cole, R. & Spiehler, V. (1989), *J. Anal. Tox.* **13** (5) 263–267.
Ritchie, L.K., Caplan, Y.H. & Park, J. (1987), *J. Anal. Tox.* **11** 205–209.
Rose, S., Cofino, J., Lee-Hearn, Wm. & Hime, G.W. (1991), Presented at *Annual Meeting of the American Academy of Forensic Sciences, Anaheim, California.*
Sasaki, Y. & Baba, S. (1988), *J. Chromatogr.* (*Biomedical Applications*), **426** 93–101.
Seno, H., Suzuki, O., Kumazawa, T. & Hattoria, H. (1991), *J. Anal. Tox.* **15** 21–5.
Shaposhnikov, A.V., Gaisin, A.A. & Titov, V.N. (1990), *Lab. Delo.* **9** 18–23.
Soo, V.A., Bergert, R.J. & Deutsch, D.G. (1986), *Clin. Chem.* **32** (2) 325–328.
Spiehler, V.R. & Reed, D. (1985), *J. Forens. Sci.* **30** (4) 1003–1011.
Street, H.V., Vycudilik, W. & Machata, G. (1977), *J. Chromatogr.* **168** 906.
Taylor, R.W., Jain, N.C. & George, M.P. (1987), *J. Anal. Tox.* **11** (5) 233–234.
Taylor, R.W., Le, S.D., Philip, S. & Jain, N.C. (1989), *J. Anal. Tox.* **13** (5) 293–295.

Van Brunt, N. (1983), *Ther. Drug Monit.* **5** 11–37.
Vereby, K. & DePace, A. (1988), *J. Chromatogr.* (*Biomedical Applications*), **427** 151–156.
Vereby, K. & DePace, A. (1989), *J. Forens. Sci.* **34** (1) 46–52.
Vinet, B. (1983), *Clin. Chem.* **29** 452–455.
Vrbanac, J.J., Braselton, W.E., Holland, J.F. & Sweeley, C.C. (1982), *J. Chromatogr.* **239** 265–276.
Whiting, J.D. & Manders, W.W. (1982), *J. Anal. Tox.* **6** 49–52.
Woodworth, J.R., Mayerson, M. & Owens, S.M. (1984) *J. Anal. Tox.* **8** 2–6.
Yamamoto, T., Takano, R., Egashira, T., Yamanaka, Y. & Teroda, M. (1989), *J. Anal. Tox.* **13** (2) 117–11.
Zhang, J.Y. & Foltz, R.L. (1990), *J. Anal. Tox.* **14** 201–205.

4

Analysis of alcohol and other volatiles

Barry K. Logan, Ph.D.
Washington State Toxicology Laboratory, Seattle, WA, USA

4.1 INTRODUCTION

The determination of ethanol in biological fluids is probably the most commonly performed forensic test in existence. The prevalence of ethanol, and its widespread social acceptance, together with a large and disparate body of rules regarding its sale and consumption, often require that its presence be shown analytically, reliably, quantitatively — and quickly. The results of this analysis are then subject to scrutiny by judges, juries, attorneys, the media and the public. The importance of the consequences which can ensue, require that all the attention and care which can be taken, must be taken.

Enzymatic methods, some of which are automated, including Enzyme Multiplied Immunoassay (EMIT), and Fluorescence Polarization Immunoassay (FPIA), are in widespread use for clinical applications, but may not be suitable for samples of the quality often encountered in forensic investigations. Classical wet chemistry methods such as colorimetric and titrimetric techniques are also still in use and can give reliable results when performed with care and adequate controls. A liquid chromatography method has recently been described for the measurement of alcohols in alcohol containing beverages (Takeuchi *et al.* 1988).

The fact that ethanol is a low boiling organic liquid, however, makes gas chromatography an ideal technique for its analysis, and this has emerged as the benchmark method for alcohol analysis. In addition to its innate compatibility with volatile compounds, gas chromatography is rapid, accurate, reproducible, and can be readily automated (Jennings 1987).

For forensic use, the reliability of a given technique must be accompanied by its use within a comprehensive program incorporating trained analysts, suitable ancillary equipment and reagents, a written protocol, regular calibration, quality control and duplicate analysis, together with appropriate measures for record keeping, sample collection, sample storage and chain of custody documentation (Smith *et al.* 1990).

Ethanol

Ethanol, or beverage alcohol, is available in everything from wine coolers to furniture polish. It is the most widely used drug in the world. Ethanol can be administered orally or rectally; absorption transdermally, or by inhalation of ethanol vapors is insignificant. Ethanol is also administered intravenously for therapeutic purposes as a competitive inhibitor of alcohol dehydrogenase, in cases of poisoning with other alcohols such as methanol, isopropanol and ethylene glycol.

Ethanol prevents the formation of toxic metabolites associated with these compounds (Ellenhorn & Barceloux 1988). Intravenous abuse of ethanol is rare but not unknown. Ethanol consumption effects a person's judgement, job performance, driving ability and health – all of which are frequent issues in criminal and civil court cases, administrative hearings and coroners' inquest. Those effects of alcohol which are considered to be desirable and account for its popularity include its ability to encourage sociability by inducing relaxation and diminishment of inhibitions. As the blood alcohol level increases, however, these can become very negative effects in an individual attempting to operate complex machinery such as a motor vehicle. Relaxation becomes inattentiveness, diminished inhibitions becomes increased risk taking, resulting in detachment from reality and confusion. In addition, alcohol induces other less pleasant side effects with no positive aspects including loss of coordination and balance, slowed reaction time, blurred or impaired vision, dizziness, and nausea, all of which obviously compound an individual's already impaired driving ability (Caplan 1982, Garriott 1988, Baselt & Cravey 1988, Ellenhorn & Barceloux 1988).

As a result of the well documented relationship between blood alcohol concentration (BAC), driving impairment and increased risk of accident involvement (Linnoila *et al.* 1986, Borkenstein *et al.* 1966), most jurisdictions have established laws against driving while intoxicated (DWI), also known as driving while under the influence (DUI), and driving under the influence of intoxicants (DUII) (Garriott 1988). These laws generally mandate a BAC level above which driving is considered unsafe and therefore illegal (Maess *et al.* 1990). However, as a result of the prevalence of both ethanol and motor vehicles, alcohol impaired driving is common. In most major towns in the United States it is estimated that on Friday and Saturday nights between 10 pm and 4 am, a majority of drivers on the road are alcohol impaired. Ethanol is frequently implicated in train, boat, bus and air crashes also, and blood ethanol levels are often an issue in under age drinking, drinking in the workplace, drinking while on probation, drinking while undergoing alcohol or drug treatment, and drinking in contravention of a court order. The average blood ethanol concentration of drivers killed in Washington State in 1991 was 0.17 g/100 ml. The range was 0.01 to 0.81 g/100 ml. Determining the alcohol content of suspected alcoholic beverages is also an issue in cases involving minors in possession of alcohol, open container in a vehicle, unlicensed establishments, illegal liquor distillation, and contravention of city or county codes on liquor possession.

In order to enforce laws that relate to any of the above types of cases, fast and accurate methods for establishing alcohol content in liquids, or blood and other body fluids are required. Blood alcohol analysis in DWI cases is the most often requested analysis in many forensic toxicology laboratories.

BAC levels are also of interest in the investigation of death. This may be death either from the acute or chronic effects of alcohol alone or in combination with other drugs, or resulting from traffic accidents where alcohol may be a contributing factor (DiMaio & DiMaio 1989). Alcohol generally will cause unconsciousness at levels of 0.35 and above in moderate drinkers, and death often results from vomiting and aspiration. Respiratory depressant effects are significant at this and higher levels and death can result from respiratory collapse with BAC levels of 0.35 and above, although blood ethanol levels as high as 0.50 g/100 ml are common, and 1.50 g/100 ml has been reported (Ellenhorn & Barceloux, 1988, O'Neill *et al.* 1984). Since the effects of alcohol are subject to the development of tolerance in heavy users, the relationship between impairment, intoxication, and potentially life threatening concentrations is indistinct, nevertheless there is very good correlation between BAC and the appearance of effects which negatively impact on driving skills.

Blood or urine alcohol determinations now also form part of many workplace drug testing programs where a person's hiring or promotion often depend in part on a negative alcohol result.

Volative liquids

Many other organic liquids have significant toxic and/or narcotic properties which can make their determination part of a forensic investigation. In addition to other alcohols such as methanol (wood alcohol) and ethylene glycol (anti-freeze), solvents such as toluene, 1,1,1, trichloroethane, freons, and the volatile components of commercial products (Table 4.1) are often abused by inhalation for their stupefying effect (Ashton 1990, Chalmers 1991). In many jurisdictions the presence of intoxicants other than ethanol can be used as evidence under DWI statutes (Maess 1988).

Many halogenated solvents with narcotic properties are used industrially as cleaning and degreasing agents. Issues often arise relating to the extent of the absorption of these chemicals in the work environment, their presence in the blood and their ability to interfere with evidential breath alcohol testing, although recent work has suggested that this is not significant (Gill *et al.* 1991a,b,c, Denney 1991a,b). The potential presence of these compounds, and investigations of these issues, require that forensic laboratories have a means of identifying all volatile chemicals in a sample when they are present, and excluding them as potential interferents when they are not.

The use of these halogenated solvents in an enclosed space with inadequate ventilation is a not uncommon industrial accident requiring post mortem investigation and toxicology (Bentur 1991, Urich *et al.* 1977).

Neither the mechanism of the deleriant effect of these chemicals, nor the mechanism of the cardiac arrythmia, coma, and death which often result following exposure, are completely understood (Casarrett & Doull 1986); however, their toxicity is well

Table 4.1. Volatile compounds in commercial products

Common name	Other names	Uses
Methanol	wood alcohol, carbinol methyl alcohol	industrial solvent, antifreeze, windshield wiper fluid, denaturant for ethanol, embalming fluid, gasoline additive
Formaldehyde	formalin, methanol	embalming fluid, manufacturing, fire gas
Ethanol	alcohol, ethyl alcohol, beverage alcohol	alcoholic beverages, cosmetic preparations, cologne, mothwash, pharmaceutical preparations, laboratory solvent, gasoline additive, paints
Acetaldehyde	ethanol	manufacturing, oxidation of ethanol
Acetone	2- propanone, dimethyl ketone	industrial solvent, paint and varnish remover, paints, pharmaceutical solvent, laboratory solvent
n-Propanol	*n*-propyl alcohol, 1-propanol	solvent for resins, laboratory reagent
Isopropanol	isopropyl alcohol 2-propanol	antifreeze, inks, denaturant for ethanol, aftershave, cologne, antiseptic wipes, manufacturing, pharmaceutical solvent
n-Butanol	*n*-butyl alcohol, 1-butanol	manufacturing, laboratory solvent, paraffin solvent
sec-Butanol	2-butanol	manufacturing, flotation agent, perfume, paints, paint remover
tert-Butanol	2-methyl-2propanol	manufacturing, flotation agent, perfume, paints, paint remover, gasoline additive
Isobutanol	2-methyl-1-propanol	paints, varnish, fruit flavoring essence manufacture
Methyl ethyl ketone	2-butanone, MEK	industrial solvent, paints and laquers
Benzene	benzol	industrial solvent, manufacturing, glues
Toluene	methyl benzene	industrial solvent, manufacturing, glues, paints and laquers, doping agents
Ethyl acetate	vinegar naphtha	fruit flavoring, fine chemical solvent, manufacturing
Methyl chloroform	1,1,1 trichloroethane	typewriter correction fluid, printing, industrial solvent
Trichloroethylene	Trilene	inhalation analgesic, typewriter correction fluid
Freon 11	trichloromonofluoromethane	refrigerant, aerosol propellant
Freon 12	dichlorodifluoromethane	refrigerant, aerosol propellant
Freon 21	dichloromonofluoromethane	refrigerant, aerosol propellant
Freon 22	monochlorodifluoromethane	refrigerant, aerosol propellant
Freon 113	trichlorofluoroethane	refrigerant, aerosol propellant
Nitrous oxide	laughing gas, dinitrogen monoxide	laboratory reagent, anaesthetic gas propellant in whipped cream machines

recognized, and their presence in blood can readily be demonstrated by gas chromatography with the appropriate sampling techniques.

Gases

Pressurized gases with narcotic properties have seen increased popularity as deleriants in recent years, particularly butane, and nitrous oxide, the latter both as a diverted anesthetic gas and a propellant for household whipped cream machines (Surudu & McGothlin 1990, Jastak 1991). Other examples appear in Table 4.1. Abuse of these gases often results in death, the investigation of which will involve the forensic toxicologists. Gases and other volatiles resulting from the burning of many synthetic materials can also cause or contribute to death. Most fire related deaths are caused by inhalation of carbon monoxide, hydrogen cyanide and nitriles rather than directly by burning (Anderson & Harland 1979).

These gases can be readily liberated from blood and tissue samples, making their analysis by gas chromatography an attractive option.

4.2 ETHANOL

4.2.1 Sample collection and sample handling

Alcohol analysis can be performed on samples from both living and dead people. Blood is the most common sample type, although at autopsy urine, vitreous humor, and spinal fluid can all be obtained. Post mortem blood samples should be drawn from peripheral or central vessels. One study examined the difference between alcohol levels in blood from different sites (Anderson & Prouty 1989), and found that the average ratio of heart blood to femoral blood alcohol for 59 cases was 0.961. In the absence of central or peripheral blood, cavity fluid samples may be collected, but quantitative results will be suspect because of the risk of contamination from gastric fluids. Vitreous humor is also a useful sample, and the typical vitreous/blood partition ratio is around 1.08. Plasma concentrations of alcohol are typically 1.10 to 1.35 (mean 1.18) times higher than the blood alcohol concentration (Payne *et al.* 1968), and a conversion can be made within these limits to report the levels as Blood Alcohol Content (BAC), the term most often used in statutes. Partition ratios for ethanol between blood and a number of other tissues have also been reported (Baselt & Cravey 1989).

Urine samples are often analyzed for alcohol content. This is not advised for forensic purposes, as there is considerable variation in the urine to blood conversion factor. As urine is almost entirely water, while blood is 80% water, the conversion factor is around 1.4, although a range of 1.1 to 2.4 has been reported (Payne *et al.* 1967, Winek & Carfagna 1987).

The issue of the rising or falling BAC is often brought up, and attempts have been made to determine the phase of the absorption curve by analyzing simultaneous blood and urine samples, as the BAC determined from a urine level typically lags the BAC by 30 minutes or so. This is however a dubious practice because of the variability in the blood/urine ratio, and usually requires an initial void with the urine and blood taken at a later time. More success with this type of estimation is possible in post mortem cases when vitreous is available. As the vitreous humor is an internal compartmentalized fluid, vitreous levels are usually lower than blood levels during absorption and higher during elimination. However, normal drinking behavior results in a peak BAC within 20–30 minutes after the last drink and the phase of the alcohol curve is rarely an issue in post mortem cases.

Samples collected from living subjects for this type of analysis should ideally be drawn into an evacuated tube designed for the purpose, containing preservatives and anticoagulants such as sodium fluoride and potassium oxalate, for reasons discussed later. This method of blood drawing is the standard practice in hospital and paramedic settings and eliminates transfer steps, reducing the risk of contamination and reducing operator contact with potentially infectious samples. Many jurisdictions insist that for a blood sample to be suitable for analysis for medicolegal purposes, the person collecting the sample be a trained doctor, nurse or phlebotomist. Oversight of this simple detail can result in a blood alcohol result being suppressed in court, regardless of how good any subsequent analysis is. If blood is being drawn for forensic purposes, ideally the arresting officer should observe the blood draw and should label and seal the tubes immediately after they are collected.

The skin at the site of drawing may be wiped with a non-alcohol containing pad before the sample is drawn. This becomes a contentious issue in many cases where either the nature of the pad is not known, or it was known to contain alcohol (this is usually isopropyl alcohol, but to many courts alcohol is alcohol). The GC method used should of course be able to make the distinction between ethanol and its homologs such as isopropyl alcohol, but this 'alcohol swab' issue can often condemn a blood alcohol analysis to suppression before the merits of the analysis itself can even be considered. Many enzymatic methods can exhibit some interference when some other alcohols are present, but the extent of the interference is not of an equimolar nature.

It can be demonstrated that drawing blood through an ethanol soaked pad can cause a 'statistically significant' increase in the blood alcohol result (McIvor & Cosbey 1990), but requires that the needle is withdrawn from the vein through the swab. This is not normal practice. Furthermore, this statistically significant increase is not meaningful, because it occurs in the third decimal place, a digit which is normally truncated from the result before it is reported (see 'calculation and reporting' below). There is emerging opinion in the medical community that the practice of swabbing an injection site with alcohol serves no useful purpose anyway and should be eliminated entirely (Peek *et al.* 1989), but it is still a common practice.

The collection tube should be approximately 75% or more filled with the sample, leaving as little air above the blood as possible to reduce the risk of any loss of volatile material. The tube must be propely sealed otherwise some alcohol can be lost through evaporation, even if stored under refrigeration.

In samples drawn from a living person, the tube used should contain both an anticoagulant, (herapin or potassium oxalate) and a preservative (sodium fluoride). The anticoagulant ensures that clotting does not take place, although even clotted samples can be analyzed satisfactorily if homogenized prior to being aliquoted for analysis. It has been postulated that samples which are partially clotted and allowed to settle, and then not shaken before analysis may result in a portion of the sample which contains a disproportionately large fraction of plasma being analyzed. Plasma concentrations of alcohol are typically 1.10 to 1.35 (mean 1.18) times higher than the blood alcohol concentration (Payne *et al.* 1968). This could theoretically result in a higher measured result. This hypothesis has never been demonstrated, however, and the extent to which clotting would have to occur to produce such an effect would certainly be obvious to the naked eye. A protocol for the analysis of any drug, including alcohol, in biological material should routinely include shaking of the sample to ensure that a representative, homogenous aliquot is being drawn. The sodium fluoride preservative is a bacterial enzyme poison, included to reduce the risk of production or degradation of ethanol by bacteria in the sample. Both production (Harper & Cory 1978) and degradation (Chang *et al.* 1984, Neuteboom & Zweipfenning 1989, Dick & Stone 1987) of ethanol in stored blood samples have been demostrated and reported. Ethanol production occurs in unpreserved samples deliberately infected with alcohol producing microorganisms or in post mortem blood samples, where blood becomes contaminated with opportunistic or gut bacteria before the samples are collected. Sterile samples such as those drawn directly into evacuated tubes have not been demonstrated to be susceptible to any bacterial induced change in alcohol concentration (Winek & Paul 1983). Loss of alcohol can occur if the tube is improperly resealed after analysis, or if stored at room temperature for extended periods.

Quality control information from the manufacturer of the blood collection tubes can often be of interest to the court in demonstrating that the preservatives were added to any particular batch of tubes. Urine samples containing high concentrations of mannitol have resulted in ethanol production even in the presence of preservatives (Jones *et al.* 1991).

Samples being collected for any type of volatiles analysis should be stored in air tight containers, and refrigerated as much as possible to prevent volatilization and loss during storage both before and after analysis. In forensic analysis this is particularly important as at some later date a defendant in a DWI case may wish to have their blood sample retested. Samples handled correctly and stored under suitable conditions can be retested for up to two years or more without any significant change in concentration (Winek & Paul 1983).

4.3 ANALYTICAL TECHNIQUE

Selection of chromatographic conditions

All the chromatographic methods discussed below for the analysis of ethanol and related compounds use flame ionization detection (FID). This 'universal' detection method provides suitable sensitivity for common concentrations of ethanol, and experience has shown it to be precise, accurate and stable in modern instruments. Mass spectrometric detection has been used for some particular applications; however, the low molecular weight of these volatiles often results in little useful mass spectrometric information. In addition, the chromatographic systems in common use have been well characterized, and the possibility of co-eluting compounds being misidentified as ethanol has all but been eliminated, making an absolute identification technique such as GCMS a very cost effective method for routine alcohol determinations.

The analysis of low boiling point solvents and gases by gas chromatography can be accomplished by introducing the sample into the chromatograph either as a liquid or a gas. Both approaches have their advantages and disadvantages.

The introduction of liquid samples may involve injection of the sample directly, or dilution or extraction prior to its injection into the heated GC injection port.

Injection of the sample as a gas involves either a liberation of the gas by reaction of the sample with a degassing agent, or heating the sample in a sealed container until the trapped vapor reaches equilibrium with the liquid sample, according to Henry's law (Jones 1983). This equilibrium or headspace vapor is then sampled in a gas tight syringe and injected into the GC.

Sample introduction considerations have significant bearing on the chromatographic conditions. Most volatile analysis is currently performed on packed columns (i.d. 2 mm or more). With direct injection analysis, where many nonvolatile artifacts from blood or tissue are being introduced directly into the GC, packed column chromatography is recommended as capillary columns can become more easily plugged, blocked or contaminated. In addition, the sensitivity of most bonded phases to water makes their use for the direct injection of water-containing samples unpractical (Jennings 1987).

When the analyte is in the gas phase, as in headspace analysis, the concentration in the gas is typically 100 to 10 000 times less than the solution or sample with which it is in equilibrium. The actual ratio depends on the analyte itself and the temperature. As a result of this difference in concentration, a correspondingly larger volume must be introduced when using a headspace technique. Sample volume in direct injection analysis is typically 1–3 μl, while in headspace analysis, samples are typically between 250 μl and 10 ml or more. There are limitations to the GC injection system which make it difficult to efficiently introduce a volume of gas this large when a GC is configured for capillary operation.

Most analyses for volatiles are done by packed column gas chromatography. The choice of the column packing is determined by the analytes of interest, and selection should be made by reference to the literature, catalogues and product information from the various vendors of chromatographic supplies, or by experimentation (Stafford & Logan 1989). The conditions for volatile analysis are usually fairly simple and can readily be adapted from the extensive literature on the subject. Most laboratories adding blood alcohol analysis to their list of services will undoubtedly choose their column, and base their method on information supplied by the chromatographic supply houses. This source of information should not be neglected, as it may be several years before results obtained on a new commercially available packing material appear in the peer reviewed literature. The various sets of chromatographic conditions discussed below are invariably suitable for the analysis of alcohols other than ethanol, and the reader is referred to the source material for more complete details.

Direct injection gas chromatographic analysis

In the early 1960s numerous methods for the analysis of alcohol in blood appeared (Jain & Cravey 1972, Cravey & Jain 1974) using both direct injection and headspace sampling techniques. It was also popular at that time to perform a clean-up step such as distillation, protein precipitation or solvent extraction of the ethanol prior to analysis. This helped to extend the lifetime of the column but introduced poorer precision. One method typical of this period which combines several of these features (Davis 1966), describes the use of dioxane which serves to precipitate proteins and act as an extraction solvent for the ethanol. The method also used an impurity in the dioxane, 2-ethoxyethanol, as an internal standard. This one-step process was favored over direct injection as it eliminated the introduction of protein and water into the gas chromatograph. Using a specialized phase consisting of Flexol 8N8, polyethyleneglycol 600 and diisodecylphthalate, analysis for the lower alcohols took up to 10 minutes. Resolution was poor between ethanol, acetone and isopropanol, and retention times were very long (Fig. 4.1).

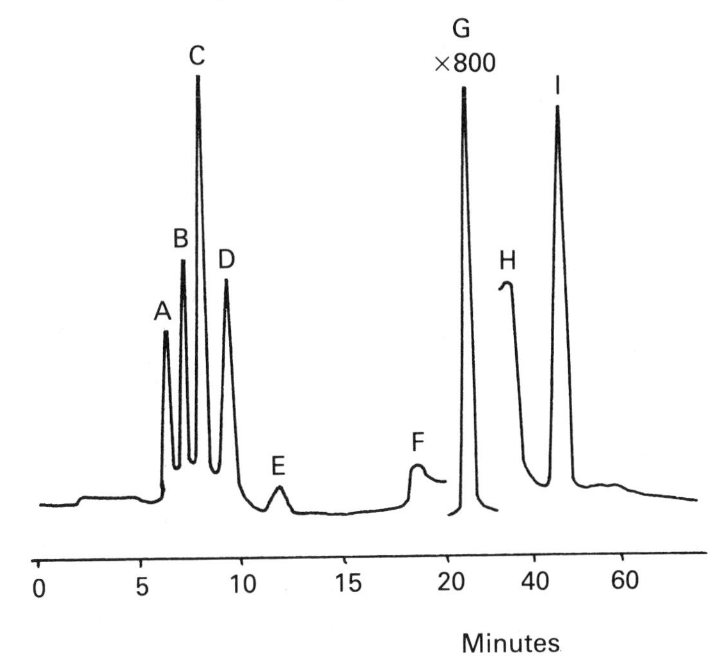

Fig. 4.1. Early gas chromatographic procedure for volatiles, using direct injection technique. Note the longer retention times, and poor resolution. Chromatography was performed on Flexol 8N8, polyethyleneglycol 600 and diisodecylphthalate (Davis 1966). (A) = methanol, 0.16 g/100 ml; (B) = acetone, 0.08 g/100 ml; (C) = ethanol, 0.16 g/100 ml; (D) = isopropanol, 0.08 g/100 ml; (E) = ethyl acetate (in the dioxane); (F) = unknown peak (in the dioxane); (G) = dioxane; (H) = methoxymethanol (in the dioxane); and (I) = ethoxyethanol, 0.46 g/100 ml dioxane.

Simple direct injection following dilution of the blood with an aqueous solution of the internal standard (Jain 1971) became more popular because the lack of any extraction step made the process more convenient for large scale analysis. Blood

(0.5 ml) and internal standard solution (50 mg/100 ml isobutanol) (0.5 ml) were mixed and 0.1 to 0.5 μl injected onto a column filled with Carbowax 20 M on a 60/80 Chromasorb W support. Blood could also be injected directly without any dilution or internal standard for qualitative analysis.

Many sets of chromatographic conditions have been reported for this technique. Successful analyses have been reported using Hallcomid (Mather & Assimos 1965), Polyethylene glycol (Curry *et al.* 1966), Porapak Q and Porapak S, Carbowax, and various combinations of the above (Finkle 1971). Many of these older methods have been reviewed extensively elsewhere (Jain & Cravey 1972).

The nature of laboratories performing blood alcohol analysis, and the availability of prepacked columns and ready made phases have changed since the time when many of these methods were developed. Most laboratories will now buy 'off the shelf' methods for many analyses including alcohol. Of the phases currently sold for alcohol analyses the most popular are undoubtedly Porapak Q/S (Giles *et al.* 1986) and Carbowax 20M (Wright 1991, Dubowski 1980).

Performing direct injection blood alcohol analysis

The recommended protocol for direct injection analysis is as follows. Clotted or inhomogeneous samples should first be homogenized by grinding or similar means. Dilution of blood samples is advised to minimize problems with clogging of microsyringes. A dilution ratio of 1:10 is recommmended. This allows for homogenization of the biological sample, helping to extend column lifetime, and encourages lysis of any remaining cells. The normal practice is to dilute the blood sample with an aqueous solution of the internal standard. This can be done by hand, but the use of an automated pipetter/diluter ensures better reproducibility, and makes the handling of large numbers of samples easier. Rinsing the pipetter/diluter between samples is good forensic practice, and helps to eliminate later questions as to the possibility of carry over. After dilution, samples should be thoroughly mixed, and capped to reduce evaporation if the sample is not to be analyzed immediately. 1–3 μl can be injected directly into the gas chromatograph. A representative chromatogram, obtained on Porapak S, is shown in Fig. 4.2.

Numerous automatic injection systems for gas chromatography exist. Their use with direct injection protocols is not recommended, however, owing to the high risk of a blocked syringe needle arising from the analysis of unextracted samples. Blocked needles can readily be cleared by heating in a flame, while exerting gentle pressure on the plunger.

Unautomated direct injection analysis is eminently suitable and a cost effective method for the analysis of small numbers of samples, but quickly becomes tedious.

Headspace analysis

According to Henry's Law, a volatile solvent in solution will partition into the air above it at a fixed ratio as a function of its concentration and temperature. By allowing volatile compounds to equilibrate with the air in a closed container and then sampling the air above the sample ('headspace'), many disadvantages associated with direct injection, such as blocked needles, nonhomogeneous aliquots, problematic

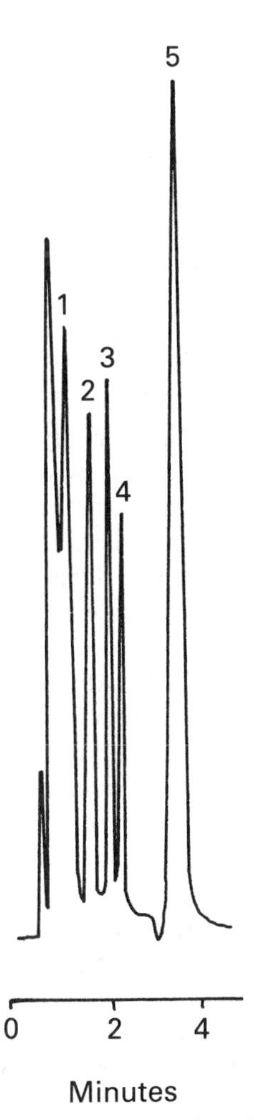

Fig. 4.2. Direct injection analysis of alcohols in whole blood. 1 Methanol, 2 ethanol, 3 acetone, 4 isopropanol, 5 methylethylketone (is). Obtained on 3 ft stainless-steel column packed with Porapak S 80/100; nitrogen carrier gas, 90°C.

automation, and column contamination can be overcome. Different volatile substances have different partition ratios, and an appropriate temperature must be selected to ensure adequate sensitivity for the analyte of interest. In automated systems the temperature of the sample loop or syringe must exceed that of the sample to avoid condensation. For ethanol, headspace analysis can be performed at room or ambient temperature (Penton 1985), however, fluctuations in room temperature in any laboratory makes this unadvisable. For practical reasons, temperatures within the range of a water or oil bath are normally used. Methods for headspace analysis of alcohol were reported as early as 1964 (Goldbaum *et al.* 1964). Temperatures ranging from 25°C to 60°C were common. Many of the same phases used for direct injection analysis were used for headspace methods also, including Flexol (Wallace & Dahl, 1966), polyethylene glycol (Machata 1962) and Carbowax (Coldwell *et al.* 1971). For similar reasons to those outlined above, the same commercially available phases have become standard for headspace analysis.

The fact that a vapor phase sample is being analyzed makes the use of wide-bore capillary or open tubular fused silica columns an additional option. In a review of three phases on 5 μm fused silica columns (Penton 1987) a 15 m methyl silicone column with a film thickness of 5 μm, gave resolution of acetaldehyde, methanol, ethanol, acetone and other C3 and C4 alcohols in 0.65 minutes or less.

Performing headspace blood alcohol analysis

The recommended protocol for headspace analysis is as follows. Dilution of blood samples is essential to ensure complete lysis of all cells and the release of any intracellular alcohol. A dilution ratio of at least 1:10 (blood/diluent) is recommended. For headspace analysis the normal practice is to dilute the blood sample with an aqueous solution of the internal standard, often also containing sodium chloride or sodium sulphate. This is in an effort to increase the volatilization of the ethanol and the internal standard, lowering the liquid/air partition ratio and improving sensitivity. This also helps to correct for any minor differences in salt concentration in the samples. The use of a high dilution factor and the addition of salt to the samples eliminates any differences between blood/air and water/air partition ratios, and thus allows the use of aqueous standard for calibration (Watts & McDonald 1987, 1990).

It is important to note that ethanol in biological samples at elevated temperatures can become oxidized, resulting in the production of acetaldehyde (Thomas *et al.* 1981).

Studies in our laboratory have shown a drop in ethanol concentration of as much as 0.02 g/100 ml when samples were incubated at 60°C for 12 hours. The change in concentration was not measurable for periods of less than 2 hours. As this does not occur with aqueous standards, ethanol concentration could be underestimated if samples are not analyzed within three to four hours of being incubated when aqueous standards are used. This underscores the importance of using a chromatographic method which will differentiate between ethanol and acetaldehyde. As methanol and acetaldehyde elute very closely on most chromatographic systems, care must be taken that acetaldehyde is separated from and distinguishable from methanol when the identity of either of these two compounds is important.

As with direct injection analysis, the use of an electronic pipetter/diluter ensures optimum reproducibility. The importance of a stable thermostatted device for heating the samples prior to injection cannot be overemphasized. The importance of stable temperature together with sample handling considerations makes the use of an automated injection device highly recommended. Injection volumes will depend on the chromatographic conditions, instrument sensitivity and sample size, and can vary from 25 μl up to 10 ml or more.

After dilution, samples should be thoroughly mixed, and capped to reduce evaporation, then placed in a constant temperature bath. Equilibration time for the sample will depend on the temperature, sample size and container and may vary from 15 minutes to usually not more than 1 hour. One must ensure that all vials or tubes are completely sealed before being placed in the bath, otherwise prolonged equilibration could result in the loss of potentially valuable sample material. Duplicate analysis is good forensic practice, and as with any automated method, tracking the identity of a vial through the analytical process should be a significant consideration. Several automatic injection systems for headspace gas chromatography exist, and vary as to available temperature range, injection volume, vial capacity, and number of samples. Important considerations in the selection of such a system should include its compatibility with existing equipment and workload.

Figure 4.3 shows a chromatogram and conditions of a headspace blood alcohol analysis performed on one of several commercially available automated headspace samplers.

As noted above, headspace analysis can be performed at room temperature although variations in room temperature may affect the precision of these results. The advantage of room temperature headspace analysis is the possibility of modifying an existing autosampler for headspace operation (Penton 1985).

Ethanol in collected breath samples

Most nondestructive evidentiary breath alcohol devices (e.g. DataMaster, Intoxilyzer, Intoximeter) can be modified to collect and save a breath sample for later analysis, by absorption of the sample onto a silica absorbant. For GC analysis the contents of this silica tube are typically emptied into a vial, diluted with an aqueous solution containing internal standard, and analyzed by headspace gas chromatography. The amount of ethanol contained in a 50 ml breath sample is, however, very small (50 ml of a 0.1 g/210 l, breath sample contains 23 μg of ethanol), and a different set of conditions from those required for blood alcohol analysis may be required. This may only involve changing the relative concentration of the internal standard and the attenuation and peak threshold on the gas chromatograph or integrator.

The practice of breath collection and reanalysis is currently performed in at least six states in the United States, as required by law or administrative rule. The perceived advantage of this practice is the availability to a defendant of a sample for retest if he or she wishes to dispute the accuracy of the infra-red breath test result, or claims the presence of other volatile substances on his breath (e.g. acetone, methylethyl ketone, isopropanol, toluene) being misidentified as alcohol. Experience in those states however has shown that discrepancies between the IR result and the subsequent gas chromatography re-analysis can be attributed to factors other than instrument imprecision. These factors include inexperienced operators, faulty sample collection, faulty sample storage, leaks in the breath test device, faulty collection tubes and also the innate analytical limitations of comparing results from the two different techniques (Farrell *et al.* 1991, Goldberger *et al.* 1986). Re-analysis of collected breath samples is therefore not recommended for forensic purposes, but rather the use of a breath test protocol which addresses through duplicate sampling, use of an external standard and specific detection for ethanol, issues of instrument malfunction, non-calibration, radiofrequency interference etc. Furthermore, the presence of significant amounts of solvents on the breath usually represents a severely toxic or potentially fatal condition, making the presence on the breath of solvents other than ethanol extremely unlikely.

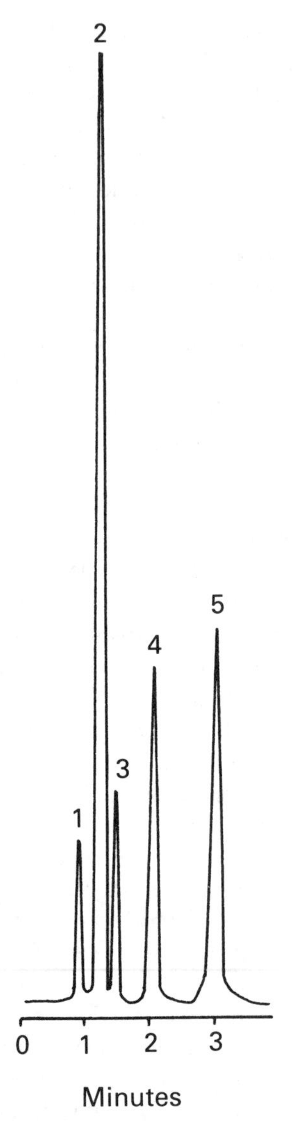

Fig. 4.3. Headspace analysis of alcohols in whole blood. 1 methanol, 2 acetone, 3 ethanol, 4 isopropanol, 5 *n*-propanol (is). Obtained on 6 ft glass column packed with 5% Carbowax 20M on Carbopak B 60/80; nitrogen carrier gas, 83°C.

This issue is continually raised in court, however, and can result in confusion or apprehension in judges and juries, unless firmly rebutted (Gill *et al.* 1991a,b,c).

Calculating and reporting results

Results obtained by using any analytical method which are destined for use in court must be absolutely defensible and must include at a minimum the following features: ensuring the identity of the sample; tracking its identity throughout the analytical process; ensuring that blanks are run periodically to demonstrate the impossibility of carry over; the periodic use of primary standards throughout any run; and the linearity, range and nature of standards which are used for calibration of the method. The use of an internal standard is essential to check chromatographic performance, and to assist in optimizing precision and accuracy. The standards used should cover a wide range which should include any critical level such as a 0.1 or 0.08 g/100 ml *per se* standard.

The precision of any method used for blood alcohol analyses should show a CV of 3% or less on replicate analyses, and should be accurate to ±5% compared to primary standards. Samples should be analyzed in duplicate, whenever any positive result is to be reported. State statutes and therefore forensic results are typically reported in grams per 100 ml of blood (g/100 ml, often g%) but one must ensure that the units are consistent with the state statute. Most clinical laboratories prefer to use units of mg/dl (0.1 g/100 ml = 100 mg/dl).

Different jurisdictions handle the reporting of results differently. Some will report the lower of the two analyses, some the higher, some the mean and others report both. Most gas chromatography methods are capable of measuring to at least the third decimal place. Since there is some uncertainty in the analytical method (±5% or better), the benefit of any doubt is usually given to the defendant by truncating the third digit rather than rounding. If a mean is to be calculated, the truncation should be done after the calculation of the three digit mean.

One of the most thorough programs in blood alcohol analysis (Jones & Schubert 1989) recommends that a method be exhaustively evaluated to determine the standard deviation on a result at the appropriate level, and that this be multiplied by 3 to obtain the 99.7% confidence interval. This is then subtracted from any analytical result prior to reporting. In that program, based on over 15 000 sets of data, it has been determined that these correction factors are 0.006 at levels of 0.05 and below, and 0.009 at levels of 0.150 and above. Where three times the standard deviation at the critical level is less than 0.01, truncation of the third digit, and reporting to two significant figures, is equally valid.

Limits of detection and limits of quantitation

Limits of detection (LOD's) and limits of quantitation (LOQ's) are terms often used without an adequate understanding of their meanings and implications. Both are statistical results based on the estimated standard deviation for a blank sample containing no alcohol. While this is on the surface, a fairly abstract concept, the calculation is easy to perform and useful in characterizing a given method. Described in detail elsewhere (Youden 1951, Jones 1991), it involves calculation of standard deviation on two or more replicates for a large number of samples (greater than 1000) and calculating the standard deviation for appropriate intervals within the span of the results. The standard deviation for individual determinations can be computed by summing the duplicate differences squared, and dividing by $n - k$, where n is the total number of measurements and k is the number of groups. The equation is $S^2 = [\Sigma(x_i - x)^2/[n - k]$. Plotting the standard deviation against the interval mean, and determining the intercept for zero BAC, gives an estimate of the standard deviation for a zero BAC result (SD_0). The limit of detection is generally considered to be 3 times the SD_0, and the limit of quantitation is 10 times the SD_0 (MacDougal *et al.* 1980, Taylor 1987). Gas chromatographic methods can have an LOD of 0.002 g/100 ml or better and an LOQ of 0.008 g/100 ml Measurement of lower levels are of course possible provided that the method is adequately characterized for that lower range.

Quality control in blood alcohol analysis

The use in courts, hearings, and inquests of the results of alcohol analysis require that certain safeguards are noted, and that documentation of results of standards, instrument certification, maintenance, external quality control are kept, and documentation is maintained to validate the results of any particular analysis. These considerations are over and above the requirements normally made of academic or commercial analyses and place. Quality control is a critical part of any analytical procedure and especially so in blood alcohol analysis. It is recommended that quality control samples be run regularly along with evidential samples. These should be obtained from a primary reference source, such as one of the clinical supply houses, or professional regulatory agencies. Long term quality assurance can be demonstrated through participation in programs whereby samples are received on a 'blind' basis and analyzed by a number of different laboratories. The results of any particular laboratory can then be based on a comparison with its peers engaged in the same type of analysis. Such programs are currently offered in the United States by among others the College of American Pathologists, and the Department of Transportation. Some states and regions will in addition license or issue permits to individuals who meet some requirements (usually a minimum educational level, use of approved method, and satisfactory performance in analyzing test samples) specifically recognizing an ability to perform blood alcohol analysis to the standards required of a forensic test.

Interpretation of results and court testimony

Following the analysis it is often the responsibility of the analyst not only to testify about the satisfactory performance of a blood alcohol test, but also to interpret his or her findings for a court, an inquest or a hearing. Statutes in most jurisdictions include a *per se* offence of DWI or DUI. This states that it is an offense in its own right (or *per se*) to drive or operate a motor vehicle with a blood alcohol content above a certain level (0.10 g/100 ml is common, 0.08 g/100 ml is becoming more so), regardless of the subject's driving performance, field sobriety tests, or tolerance to the effects of alcohol. If the 0.10 or 0.08 level can be proved, it creates what is called an *irrebuttable presumption of guilt*, and (in theory at least) no other evidence is required in order to obtain a conviction. This has made the validity of the blood alcohol test a key issue for a defendant to attack in court, and blood alcohol tests are now regularly challenged. One major advantage of blood alcohol testing over breath testing is the availability of a stable and re-testable specimen which the defendant may have analyzed by his or her own expert if he or she wishes to dispute a set of test results.

The physiology and effects of alcohol are dealt with extensively elsewhere (Ellenhorn & Barceloux 1988, Garriott 1988, Caplan 1982), but other issues refered to above which often arise during testimony and with which an analyst should be familiar, include the possibility of the production of alcohol in post mortem material, contamination from swabs, the effects of the presence of anticoagulants or preservatives, the significance of the presence of other alcohols, the principle and specificity of the method, quality control procedures and maintenance records for the instruments, the existence of a written protocol and how closely it was adhered to, the reliability of the method, its general acceptance by the scientific community, quality control programs, participation in accreditation schemes, the training and qualifications of the analyst, and a host of other issues, both relevant and irrelevant.

4.4 OTHER VOLATILE CHEMICALS

Introduction

Although alcohol represents the most common analyte in forensic laboratories, many other volatiles and solvents are readily available and may feature in an investigation either of intoxication or post mortem toxicology. Exposure can result either from deliberate abuse of these materials for their narcotic effects or from industrial exposure (Oliver 1982, Lush *et al.* 1982, McHugh 1987, Uehori *et al.* 1987, Bentur & Koren 1991, Gill *et al.* 1988). Table 1 lists those in widespread use. The materials most commonly encountered are methanol, isopropyl alcohol, acetone, toluene, methylethyl ketone, and a variety of chlorinated/fluorinated hydrocarbons or freons.

Gas chromatography of the non-halogenated volatiles can typically be performed on the same chromatographic systems as ethanol, with modifications of temperature (Gill *et al.* 1988, Jones 1988). Acetone, methanol, propanol and butanol isomers will chromatograph readily on almost all the direct injection and headspace systems discussed above, and the reader is referred to these for more complete details.

The analysis of chlorinated/fluorinated hydrocarbons is more problematic because of the poor response of flame ionization detectors to these compounds. This often requires the use of electron capture detectors, which are extremely sensitive to halogenated compounds. This approach has been most successful for quantitation of these compounds (Urich *et al.* 1977); however, most electron capture detectors are configured for use with capillary columns, and are not commonly set up for volatiles analysis.

An alternative qualitative method uses GC with mass spectrometry. One advantage of this method is that a mass spectral fingerprint can allow discrimination between many similar freons. This can be performed by adding a sample to a headspace vial, heating to about 50°C, and injecting approximately 1 ml of headspace directly into most GCMS systems. The detector should be turned on prior to injection, to allow collection of the 'air' peak which will contain any volatiles present. Examination of the air peak for characteristic ions can confirm the presence and identity of halogenated hydrocarbons. Poor peak shape and the difficulty of manipulating these volatile materials make the preparation of standards and any subsequent quantitation difficult. Nitrous oxide, an inhalation anesthetic and deliriant, is frequently abused by both medical and dental professionals who have access to it, and also by people abusing the small quantities sold as propellants from whipped cream dispensers. Nitrous oxide is rapidly metabolized, but can be detected in blood, tracheal air or tissue by using a headspace sampling technique followed by analysis on a gas chromatograph or GCMS system (Heusler 1982). This system is also suitable for many of the halogenated medical anesthetic gases.

4.5 CONCLUSIONS

Gas chromatography is ideally suited to the analysis of volatile compounds in biological materials. Blood alcohol analysis is the most commonly requested analysis performed by most forensic laboratories, and the dedication of an instrument specifically for this application is usually warranted. In addition, the ability of most methods to simultaneously measure specifically, quantitatively and accurately, methanol, isopropanol and acetone makes gas chromatography the preferred method. The ability to automate the procedure makes headspace GC eminently suitable for large throughput forensic laboratories.

Additional considerations required of a forensic laboratory include thorough documentation, rigorous quality control, complete chain of custody records, consistent reporting of results, and a simple explanation of the method and reasoned interpretation of results for lay people.

REFERENCES

Ashton, C.H. (1990), Solvent abuse. *British Medical Journal*, **300** 135–136.

Anderson, R.A. & Harland, W.A. (1979), The analysis of volatiles in blood from fire fatalities; in *Forensic Toxicology. Proceedings of the European Meeting of the International Association of Forensic Toxicologists*. 279–291. (Oliver, J.S. ed.).

Anderson, W.H. & Prouty, R.W. (1989), Postmortem distribution of drugs. In *Advances in Analytical Toxicology*. **II**, Yearbook Medical Publishers, 70–103.

Baselt, R.C. & Cravey, R.H. (1989), *Disposition of toxic drugs and chemicals in man*. 3rd ed. Year Book Medical Publishers.

Bentur, Y. & Koren, G. (1991), The three most common occupational exposures reported by pregnant women: an update. *American Journal of Obstetrics and Gynecology*, **165** (2) 429–437.

Borkenstein, R.F., Crowther, R.F., Shumate, R.P., Ziel, W.B. & Zylman, R. (1974), The role of the drinking driver in traffic accidents (the Grand Rapids study). *Blutalkohol*, **11** (supp. 1) 1–132.

Caplan, Y.H. (1982), The determination of alcohol in blood and breath. In *Forensic Science Handbook*, Prentice-Hall. Saferstein, R. (ed.).

Casarrett and Doull's Toxicology, Klaassen, C.D., Amdur, M.O. & Doull, J. (1986), 3rd edn. Macmillan.

Chalmers, E.M. (1991), Volatile substance abuse. *Medical Journal of Australia*, **154** 269–274.

Chang, R.B., Smith, W.A., Walkin, E. & Reynolds, P.C. (1984), The stability of ethyl alcohol in forensic blood specimens. *Journal of Analytical Toxicology*, **8** (2) 66–67.

Coldwell, B.B., Solmonraj, G., Trenholm, H.L. & Wiberg, H.S. (1971), *Clinical Toxicology*, **13** 374–379.

Cory, J.E.L. (1978), Possible sources of ethanol ante- and postmortem: its relationship to the biochemistry and microbiology of decomposition. *Journal of Applied Bacteriology*, **44** 1–56.

Cravey, R.H. & Jain, N.C. (1974), Current status of blood alcohol methods. *Journal of Chromatographic Science*, **12** 209–213.

Curry, A.S., Walker, G.W. & Simpson, G.S. (1966), Determination of ethanol in blood by gas chromatography. *Analyst*, **91** 742–746.

Davis, R.A. (1966), The determination of ethanol in blood or tissue by gas chromatography. *Journal of Forensic Sciences*, **11** (2) 205–213.

DeMaster, E.G., Redfern, B., Weir, E.K., Pierpont, G.L. & Crouse, L.J. (1983), Elimination of artifactual acetaldehyde in the measurement of human blood acetaldehyde by the use of polyethylene glycol and sodium azide: Normal blood acetaldehyde levels in the dog and human after ethanol. *Alcoholism: Clinical and Experimental Research*, **7** (4) 436–441.

Denney, R.C. (1991a), Inhalation and absorption of toluene and xylene in a working environment. Bulletin of the International Association of Forensic Toxicologists, April 1991, **21** (2) 27–29.

Denney, R.C. (1991b), A study of methanol, toluene and xylene absorption and inhalation from paint spraying. Bulletin of the International Association of Forensic Toxicologists, April 1991, **21** (2) 25–27.

Dick, G.L. & Stone, H.M. (1987), Alcohol loss arising from microbial contamination of drivers' blood specimens. *Forensic Science International*, **34** (1–2) 17–27.

DiMaio, D.J. & DiMaio, J.M. (1989), *Forensic pathology*, Elsevier.

Dubowski, K.M. (1980), Alcohol determination in the clinical laboratory. *American Journal of Clinical Pathology*, **74** 747–750.

Ellenhorn, M.J. & Barceleoux, D.G. (1988), *Medical toxicology. Diagnosis and treatment of human poisoning*. New York. Elsevier.

Farrell, L., Zettl, J.R. & Davia, M.T. (1991), A statewide evidential delayed breath alcohol program – what can you realistically expect? Presented at the *American Academy of Forensic Sciences.*

Finkle, B.S. (1971), in *Manual of analytical toxicology*. Cleveland: CRC Press, (Sunshine, I. (ed.)).

Garriott, J.C. (ed.) (1988), *Medicolegal aspect of alcohol determination in biological specimens*: PSG.

Giles, H.G., Meggiorini, S. & Vidins, E.I. (1986), Semiautomated analysis of ethanol and acetate in human plasma by headspace gas chromatography. *Canadian Journal of Physiology and Pharmacology*, **64** (6) 717–719.

Gill, R., Hatchett, S.E., Ossleton, M.D., Wilson, H.K. & Ramsey, J.D. (1988), Sample handling and storage for the quantitative analysis of volatile compounds in blood: the determination of toluene by headspace gas chromatography. *Journal of Analytical Toxicology*, **12** 141–155.

Gill, R., Hatchett, S.E., Broster, C.G., Ossleton, M.D., Ramsey, J.D., Wilson, H.K. & Wilcox, A.H. (1991a), The response of evidential breath testing instruments with subjects exposed to organic solvents and gases I. Toluene, 1,1,1 trichloroethane and butane. *Medicine, Science and the Law*, **31** (3) 187–200.

Gill, R., Warner, H.E., Broster, C.G., Ossleton, M.D., Ramsey, J.D., Wilson, H.K. & Wilcox, A.H. (1991b), The response of evidential breath testing instruments with subjects exposed to organic solvents and gases II. White spirit and nonane. *Medicine, Science and the Law*, **31** (3) 201–213.

Gill, R., Ossleton, M.D., Broad, J.E. & Ramsey, J.D. (1991c), The response of evidential breath testing instruments with subjects exposed to organic solvents and gases II. White spirit exposure during domestic painting. *Medicine, Science and the Law*, **31** (3) 214–220.

Goldbaum, L.R., Domanski, T.J. & Schloegal, E.L. (1964), Analysis of biological specimens for volatile compounds by gas chromatography. *Journal of Forensic Sciences*, **9** 63–71.

Goldberger, B.A., Caplan, Y.H. & Zettl, J.R. (1986), A long term experience with breath ethanol collection employing silica gel. *Journal of Analytical Toxicology*, **10** 194–197.

Harper, D.R. & Corry, J.E.L. (1988), Collection and storage of specimens for alcohol analysis; in *Medicolegal aspects of alcohol determination in biological specimens*. PSG, (Garriott, J.C. (ed.)).

Heusler, H. (1985), Quantitative analysis of common anaesthetic agents. *Journal of Chromatography*, **340** 273–319.

Jain, N. (1971), Direct blood injection method for gas chromatographic determination of alcohols and other volatile compounds. *Clinical Chemistry*, **17** (2) 82–85.

Jain, N.C. & Cravey, R.H. (1972), Analysis of alcohol. II. A review of gas chromatographic methods. *Journal of Chromatographic Science*, **10** 263–267.

Jain, N. & Cravey, R.C. (1974), A review of breath alcohol methods. *Journal of Chromatographic Science*, **12** 214–218.

Jastak, J.T. (1991), Nitrous oxide and its abuse. *Journal of the American Dental Association*, **122** (2) 48–52.

Jennings, W. (1987), *Analytical gas chromatography*. Academic Press Orlando, Fl.

Jones, A.W. (1988), Breath acetone concentrations in fasting male volunteers: Further studies and effect of alcohol administration. *Journal of Analytical Toxicology*, **12** 75–79.

Jones, A.W. (1983), Determination of liquid/air partition coefficients for dilute solutions of ethanol in water, whole blood and plasma. *Journal of Analytical Toxicology*, **7** (4) 193–197.

Jones, A.W. (1991), Limits of detection and quantitation of ethanol in specimens of whole blood from drinking drivers analyzed by headspace gas chromatography. *Journal of Forensic Sciences*, **36** (5) 1277–1279.

Jones, A.W., Anderson, R., Sakshaug, J. & Morland, J. (1991a), Possible formation of ethanol in postmortem blood specimens after antemortem treatment with mannitol. *Journal of Analytical Toxicology*, **15** 157–158.

Jones, A.W., Nilsson, L., Gladh, S.Å., Karlsson, K. & Beck-Friis, J. (1991b), 2,3-Butanediol in plasma from an alcoholic mistakenly identified as ethylene glycol by gas chromatographic analysis. *Clinical Chemistry*, **37** (8) 1453–1455.

Jones, A.W. & Schubert, J. (1989), Computer aided headspace gas chromatography applied to blood alcohol analysis: importance of online process control. *Journal of Forensic Sciences*, **34** 1116–1127.

Linnoila, M., Stapleton, J., Lister, R., Guthrie, S. & Eckardt, M. (1986), Effects of alcohol on accident risk. *Pathologist*, August, 36–41.

Lush, M., Oliver, J.S. & Watson, J.M. (1980), The analysis of blood in cases of suspected solvent abuse, with a review of results during the period October 1977 to July 1979. In *Forensic Toxicology. Proceedings of the European Meeting of the International Association of Forensic Toxicologists*. 304–313. (Oliver, J.S. ed.).

MacDougal, D. *et al.* (1980), Guidelines for data acquisition and data quality evaluation in environmental chemistry. *Analytical Chemistry*, **52** 2242–2249.

Machata, G. (1962), Die routineuntersuchung der blutalkoholkonzentration mit der gas-chromatographie. *Mikrochimica Acta*, **4** 691.

Maess, J.E. (ed.) (1990), *Critical issues: drunk driving prosecutions. Annotations from the ALR system*, Lawyers cooperative publishing.

Mather, A. & Assimos, A. (1965), Evaluation of gas-liquid chromatography assays for blood volatiles. *Clinical Chemistry*, **11** 1023–1035.

McHugh, M.J. (1987), The abuse of volatile substances. *Pediatric Clinics in North America*, **34** (2) 333–340.

McIvor, R.A. & Cosbey, S.H. (1990), Effect of using alcoholic and non-alcoholic skin cleansing swabs when sampling blood for alcohol estimation using gas chromatography. *British Journal of Clinical Practice*, **44** (6) 235–236.

Neuteboom, W. & Zweipfenning, P.G. (1989), The stability of the alcohol concentration in urine specimens. *Journal of Analytical Toxicology*, **13** (3) 141–143.

Oliver, J.S. (1982), The analytical diagnosis of solvent abuse. *Human Toxicology*, **1** 293–297.

O'Neill, S., Tipton, K.F., Prichard, J.S. & Quinlan, A. (1984), Survival after high blood alcohol levels. *Archives of Internal Medicine*, **144** 641–642.

Payne, J.P., Foster, D.V. & Hill, D.W. (1967), Observations on interpretation of blood alcohol levels derived from analysis of urine. *British Medical Journal*, **3** 819–823.

Payne, J.P., Hill, D.W. & Wood, D.G.L. (1968), Distribution of ethanol plasma and erythrocytes in whole blood. *Nature*, **217** 963–964.

Peek, G.J.P., Keating, J.W., Ward, R.J. & Peters, T.J. (1989), Alcohol swabs and venepuncture. *The Lancet*, June 17, 138.

Penton, Z. (1987), Gas chromatographic determination of ethanol in blood with 0.53 mm fused silica open tubular columns. *Clinical Chemistry*, **33** (11) 2094–2095.

Penton, Z. (1985), Headspace measurement of ethanol in blood by gas chromatography with a modified autosampler. *Clinical Chemistry*, **31** (3) 439–41.

Smith, M.L., Bronner, W.E, Shimomura, E.T., Levine, B.S. & Froede, R.C. (1990), Quality assurance in drug testing laboratories. *Clinics in Laboratory Medicine*, **10** (3) 503–516.

Stafford, D.T. & Logan, B.K. (1990), Information resources useful in forensic toxicology. *Fundamental and applied toxicology*, **15** 411–419.

Sullivan, J.B., Hauptman, M. & Bronstein, A.C. (1987), Lack of observable intoxication in humans with high plasma alcohol concentrations. *Journal of Forensic Sciences*, **32** 1660–1665.

Suruda, A.J. & McGlothlin, J.D. (1990), Fatal abuse of nitrous oxide in the workplace. *Journal of Occupational Medicine*, **32** (8) 682–684.

Takeuchi, T., Murase, K. & Ishii, D. (1986), Determination of alcohol in alcoholic beverages by high performance liquid chromatography with indirect photometric detection. *Journal of Chromatography*, **445** 139–144.

Taylor, J.K. (1987), *Quality assurance of chemical measurements*, Chelsea, Mich., Lewis Publishers Inc., 79–83.

Thomas, M., Lim, C.K. & Peters, T.J. (1981), Assaying acetaldehyde in biological fluids. *The Lancet*, September 5, **2** 530–531.

Uehori, R., Nagata, T., Kimura, K., Kudo, K. & Noda, M. (1987), Screening of volatile compounds present in human blood using retention indices in gas chromatography. *Journal of Chromatography*, **411** 251–257.

Urich, R.W., Wittenberg, P.H., Bowerman, D.L., Levisky, J.A. & Pflug, J.L. (1977), Electron impact mass spectrometric detection of freon in biological specimens. *Journal of Forensic Sciences*, **22** 34–39.

Wallace, J.E. & Dahl, E.V. (1966), Rapid vapor phase method for determining ethanol in blood and urine by gas chromatography. *American Journal of Clinical Pathology*, **46** (1) 152–154.

Watts, M.T. & McDonald, O.L. (1987), The effect of biologic specimens on the gas chromatography headspace analysis of ethanol and other volatile compounds. *American Journal of Clinical Pathology*, **87** 79–85.

Watts, M.T. & McDonald, O.L. (1990), The effect of sodium chloride concentration, water content, and protein on the gas chromatographic headspace analysis of ethanol. *American Journal of Clinical Pathology*, **93** 357–362.

Winek, C.L. & Carfagna, M. (1987), Comparison of plasma, serum, and whole blood ethanol concentrations. *Journal of Analytical Toxicology*, **11** 267–268.

Winek, C.L. & Paul, L.J. (1983), Effect of short-term storage conditions on alcohol concentrations in blood from living human subjects. *Clinical Chemistry*, **29** (11) 1959–1960.

Wright, J.W. (1991), Alcohol and the laboratory in the United Kingdom. *Annals of Clinical Biochemistry*, **28** (3) 212–217.

Youden, W.J. (1951), *Statistical methods for chemists*, London, Chapman and Hall.

Zink, P. & Reinhardt, G. (1988), The course of the blood alcohol curve during and after consumption of alcohol in large quantities. Long time studies on human volunteers. In *Biomedical and social aspects of alcohol and alcoholism*. Kuriyama, H. (ed.) Elsevier Science Publishers, 623–628.

5

Gas chromatography in arson and explosives analysis

Mary Lou Fultz, Ph.D.
Bureau of Alcohol, Tobacco and Firearms, Rockville, MD, USA

John D. DeHaan, B.S.
Bureau of Forensic Services, California Criminalistics Institute, Sacramento, CA, USA

5.1 OUTLINE

Gas chromatography has found widespread use and is the method of choice in forensic laboratories in the analysis of fire debris. The technique is less widely used in the analysis of explosives, but has found some accepted applications. This chapter will review the use of gas chromatography in the forensic analysis of fire and explosion debris. General analysis conditions as well as specific applications are covered. The information presented, along with the references cited, provides the reader with a broad overview of gas chromatographic analysis of arson and explosives samples.

5.2 ARSON ACCELERANT DETECTION ANALYSIS

5.2.1 Introduction

A deliberately set fire is one of man's oldest weapons against another. In addition to the physical threat of death or injury, a fire denies shelter, protection, food, or income to the victim. Incendiary fires often involve the use of some material to insure ignition, accelerate the rate of fire spread, or promote the spread of fire. Any such material can be defined as an accelerant (Lentini *et al.* 1989)§. Today, the most commonly used and detected are commercially available flammable or combustible liquids.

§For a comprehensive listing of terms used in fire debris analysis, see 'Glossary of Terms Related to Chemical and Instrumental Analysis of Fire Debris' in *Fire and Arson Investigator*, **40**, (2), 25–34, 1989.

Examples of such products include, but are not limited to, automotive gasoline, kerosene, paint thinners, charcoal lighter fluids, alcohols, ketones, mineral spirits, and fuel oils. These products are inexpensive, readily available, and effective in aiding fire spread.

Fire investigators first seek the origin of the fire, the point or area where it began, and then look for suitable sources of ignition. There may be indicators at the scene that a fire was deliberately set, such as timing or delay devices, or even the absence of a single identifiable point of origin. One primary indicator is that the severity of damage or speed of spread is not explainable by the ignition of ordinary combustibles (paper, wood, rubber, plastic). Witnesses may report that the building was nearly empty of likely fuels or that it 'exploded' in flame. Investigators look for patterns that flammable or combustible liquids sometimes leave when they are burned on surfaces – areas of discoloration or severe damage – which do not match known fuels or ventilation conditions. Areas protected from the heat of the fire may contain residues of the original liquid. Petroleum products are not water-soluble and may be sealed into absorbent substrates (wood, soil, concrete) by the water from fire fighting. While non-petroleum accelerants (alcohols and ketones) are water soluble and thus may be lost in an excess of water, they are sometimes protected by simple absorption into wood, upholstery materials, or plastics. Detection and identification of the accelerant present may provide the investigator with incontrovertible proof that the fire was incendiary or even provide investigative leads to the identity of the perpetrator. Debris recovered from the scene, however, is usually wet, burned or partially burned carpet, pad, wood, tile, upholstery, or paint, any one of which can contribute its own volatile pyrolysis products. The task facing the forensic laboratory analyst is the recovery and isolation of the accelerant from the fire debris, followed by identification.

The recovery and identification of flammable or combustible liquids using gas chromatography is discussed in the following sections. Because identification of these products requires an understanding of the manufacturing processes used to make them as well as fire behavior, an overview of each topic will be presented first. The implications of refinery processes on chromatographic interpretation will be discussed, along with sample preparation methods and the chromatographic analysis itself.

5.2.2 Fire behavior

It is helpful to understand the basic development of both normal and accelerated fires so that the analyst can better appreciate the effects on the evidence being analyzed. Although fire is a complex interaction of air, fuel, and heat, fires grow according to fairly predictable pathways or models. These are defined by the size of the room, arrangement and nature of the fuels present, and the nature of ventilation. These factors can be used to predict the speed, intensity, and direction of fire growth and, thereby, help the investigator recognize an accelerated fire. A normal fire begins with a single source of ignition and the first fuel it contacts. If that fuel can be raised to its ignition temperature, the flame will spread. Convection of hot gases carries the heat upward, spreading the fire vertically if there is more fuel above. The flame spreads outward more slowly by radiation and conduction. As more fuel is involved, the convective plume grows taller, bringing it into contact with more and more fuel and creating a layer of hot gases at the ceiling. This layer will lose some of its heat by conduction into walls and ceilings but will tend to grow deeper and hotter as the fire progresses. Ignitable materials close to the ceiling may be ignited by the hot gases and fall down, creating more sources of ignition. If the fire is in a fuel which creates a hot smoky flame, such as plastic, the ceiling layer will contain a rich source of fuel in the form of soot, gases, and aerosols of pyrolysis products of the fuels involved. Once that mixture of fuels reaches its ignition temperature, the layer will ignite, spreading fire across the top of the room. If those gases escape the door openings, the fire will be spread into adjoining rooms. The hot gas layer may become energetic enough to produce sufficient radiant heat to ignite other fuels in the room. Upon ignition of all other fuels the room is said to have undergone flashover. Until flashover, temperatures at floor level are moderate, and the oxygen content may be near that of normal air, 21%. Upon flashover, the temperatures at floor level reach and sometimes exceed those at ceiling level (*ca* 500–700°C) and the oxygen content can drop to near zero. Materials which were decomposing or burning by normal combustion in an oxygen-rich atmosphere are now undergoing true pyrolysis, with little or no oxygen. The time required for a room to proceed from ignition of normal fuels to flashover can vary from two minutes to two hours or more, depending on the size and shape of the room, nature of fuels, ventilation, and insulation. Some rooms never go to flashover and do not burn as quickly or as completely as a result.

By contrast, an accelerated fire produces a skewed balance between fuel, ventilation, heat, and the time required for the fire to develop. A great deal of fuel is available for ready, if not instantaneous, ignition, and a great deal of heat is released very quickly. The expansion of the hot waste gases produced may overpressure the room, causing failure of windows, doors, and even walls and ceilings. The available oxygen may be used up, leaving an excess of fuel. High temperatures are produced throughout the space, from floor to ceiling, sometimes causing generalized melting, scorching, or charring damage of ceilings, carpets, furnishings, and wall-coverings. Often, however, there is not sufficient fuel and/or air to sustain the intensity and the flames die out quickly, sometimes so quickly that only the lightest, most easily ignitable solid fuels are ignited. Pools of flammable liquid may produce sustained flames, but the liquid often protects the surface underneath from the most intense flames. The flame temperatures produced by flammable liquid pools burning in air are not significantly higher than those produced by ordinary combustibles (Henderson & Lightsey) but their distribution across the lower portions of the room are in distinct contrast to the higher ceiling temperatures produced by ordinary fires. Some plastics produce the highest temperatures found in room fires, much higher than those from a pool of gasoline. Because of the intense, but often very brief, exposure to very high temperatures in a flammable liquid fire, many of the fuels in the room are never heated to their ignition temperatures; flashover is not a guaranteed result. For all these reasons, the debris from accelerated fires may be subjected to substantially different temperature and ventilation conditions than that from most ordinary combustible fires. When the pyrolysis products from debris are being evaluated by the lab analyst, it is important to know whether the fire has gone to flashover, and to have an appreciation for the effects of both flashover and accelerants.

Since there is not sufficient space to treat the topic of fire behavior and dynamics in depth here, the reader desiring more extensive background should consult specialist literature (DeHaan 1991, Drysdale 1985).

5.2.3 Collection and packaging of evidence

Residues of flammable and combustible liquid accelerants can range from bulk amounts standing in puddles to minute traces absorbed into partially burned debris. Bulk amounts can be collected directly with a clean syringe or eyedropper, or can be absorbed into clean cotton wool, gauze pads, diapers, or even toilet tissue or sanitary napkins. Smaller quantities can be asborbed into diatomaceous earth, lime (calcium carbonate), cat litter, flour, or any other clean, finely divided, inert material. A study using lime and flour (Tontarski 1985) showed that both gasoline and diesel fuel could be recovered from a concrete surface up to five days after the concrete was exposed to the fuel. The cleanliness of any absorbent is critical, especially when only minute traces of flammable liquid are present. Many common products contain oils to control dust or odors, so materials should be chosen with care. A control sample of any absorbent used should be sealed into a separate container for laboratory testing. It has recently been suggested that activated charcoal can be placed under a tent of aluminium foil on concrete or similar surfaces, warmed with a heat lamp for an hour, and then stored in an appropriate container. Some investigators have tried using spun polypropylene material used to contain petroleum spills for the recovery of liquid accelerants residues. A recent study (Fultz *et al.* 1991) showed that, while the material is efficient in absorbing petroleum products, it can contain contamination from manufacture or storage that will interfere with chromatographic interpretation. Material stored in a warehouse was found to contain trace amounts of gasoline. This material is not recommended for the recovery of liquid residues at fire scenes.

Because of the volatility of flammable or combustible liquids, they must be packaged in vapor tight containers for transmittal to the laboratory. Containers must not contain any materials that could contaminate the sample. Because of the sensitivity of some sampling techniques, the cleanliness of the container must be established by testing with the most sensitive technique available, such as charcoal adsorption/elution. Every laboratory should periodically check containers used by its client investigators for any background contamination that could interfere with chromatographic analysis of samples.

Clean, new, metal paint cans with friction lids, which come in a variety of sizes, are commonly used to package fire debris. They are sturdy, vapor tight, and resealable. Cans containing wet debris may rust over time. Lined paint cans used for water-based paints are available which reduce the problem of cans rusting through before analysis. One study has shown that cans lined with epoxy resins do not provide background contamination (DeHaan 1978). Some analysts have reported finding traces of the thinners used in some coating varnishes. Any container should be batch-tested by storing for a time with a quantity of activated charcoal, followed by carbon disulfide extraction, and gas chromatographic analysis of the eluent. Long-term storage of a year or more of dry debris may result in the loss of traces of volatiles even if the container remains intact. It is suspected that the char debris causes chemical degradation or irreversible adsorption. Residues from incendiary mixtures, including sulfur, chlorine, sodium, or potassium, are particularly prone to attack metal cans, even coated ones. Cans have been seen to fail in a matter of a few days when containing wet incendiary residues.

Glass jars can also be used to package fire debris (Hurteau 1973). Advantages are the visibility of the sample and the economy and ready availability of the containers. One problem with glass jars is that they are breakable. A second problem is that the rubber sealant in the lids may be destroyed by solvent vapors in the sample or by contact with liquid accelerant if present in bulk. With the loss of the rubber seal, the container is no longer vapor tight. It has been suggested that a seal of aluminium foil be placed under the lid to eliminate contact between the rubber seal and either the vapor or liquid residues of accelerants.

Schwanebeck (1984) measured the vapor tight properties of a variety of containers by placing a sample of gasoline in the container, and placing the container inside a sealed bell jar. The vapor within the bell jar was periodically tested for gasoline. Glass reagent bottles lost about 5% vapor/month, while polyethylene bottles lost about 5% vapor/day. A ground glass stoppered flask showed no loss of vapor in a month. Jam jars and coffee tins lost more than 5% vapor/day, while mason jars lost about 3% vapor/month. Of the bags tested, polyethylene showed very high losses, while thin polyester performed marginally. Nylon bags exhibited poor retention of gasoline. Bonded polyester/polyethylene bags showed the best retention of the bags studied.

Because cans and jars are bulky and often do not come in sizes suitable for larger pieces of evidence, the use of plastic bags has been investigated for packaging fire debris. Polyethylene containers are permeable to hydrocarbon vapors and not suitable for fire debris packaging (Tontarski 1983). Bags made of nylon film have been shown to be effective in retaining volatile compounds but can be difficult to seal tightly (DeHaan 1978). Such bags have been in use by the Metropolitan Police Forensive Science Laboratory since the early 1970s and have given good service as long as some care is taken to twist, fold, and tape the top seal. Recently, one US manufacturer has released nylon bags which are heat-sealable, overcoming the sealing problem.

In 1981, heat-sealable polyester/polyolefin bags sold by KAPAK™ were evaluated for suitability in packaging fire debris (DeHaan & Skalsky 1981). The bags were effective in retaining volatiles for most commonly encountered flammable or combustible liquids. However, in 1985, the manufacturer of the filmstock changed processes, and bags manufactured after 1985 were found to contain traces of a medium petroleum distillate (Dietz & Mann 1988). Attempts to rectify the problem were unsuccessful until recently. Now Kapak has found a new source of filmstock. These new bags were evaluated and found to be free of contamination (Kinard & Midkiff 1990). Release of the new bags was scheduled for 1991. This problem only underscores the importance of laboratories periodically evaluating containers used for fire debris for background contamination. The extraction and isolation methods in use today are so much more sensitive than previous methods (see Section 5.2.3) that even minute traces present risk of misidentification. The same cautions apply to the scene investigators, their tools, clothing, and evidence storage facilities. Debris can yield false positive results if contaminated by dirty tools, gloves, or even fleeting exposure to an open gasoline tank filler.

5.2.4 Petroleum product characterization

5.2.4.1 Refinery processes

Petroleum products originate from crude oils, which contain hundreds of organic compounds made up of carbon and hydrogen, with lesser amounts of sulfur, nitrogen, oxygen, and traces of some metals (Dyroff 1989). The variety of compounds is complex, but the majority are paraffinic, naphthenic and aromatic. The refined products made from crude oil also contain hundreds of these compounds. Because of the volatility and complexity of these products, gas chromatography is the analytical method of choice for their detection and characterization when recovered from fire debris (Forensic Science & Engineering Committee 1988).

The gas chromatographic patterns obtained from petroleum products are a result of their crude oil source and manufacturing process. The forensic analyst must have a basic understanding of these processes to correctly interpret chromatographic patterns. While it is beyond the scope of this chapter to thoroughly review refinery processes, several sources are available for review. In 1979, Thornton & Fukayama wrote a thorough review on the implications of refining operations to the characterization and analysis of petroleum products. The two-part series covers the basics of the refining of crude oil to produce marketable products and the forensic implications of these proesses. Other informative sources are the chapter on petroleum refinery processes in the Kirk/Othmer *Encyclopedia of chemical technology* (1982), and a review by King in *Occupational Medicine* (1988) on the composition, analysis and processing of petroleum.

Basically, the refinery process involves four different procedures: separation, conversion, treatment, and blending. In the separation process, compounds are physically separated and no new products are formed. In conversion, chemical changes occur by rearranging molecular structure and breaking down complex hydrocarbons to simpler molecules. Treatment processes are used to make a product conform to a specific need, such as color, odor, or chemical stability. Treatment processes may be done at the end of the refinery processes or during intermediate stops as necessary. The fourth process, blending, involves the mixing of fuel products with other oils or derivatives to provide a higher quality product. One example is gasoline, where aromatic compounds are blended with a distillate cut to obtain a product that is suitable for use in today's automotive engines. Then antioxidants, anticorrosives, and other special additives are included.

5.2.4.2 Chromatographic characterization

5.2.4.2.1 *Petroleum product classification scheme*

The variety of commercially available petroleum based products is quite large. Only those of suitably low flashpoint and volatility are used as accelerants, so the heaviest products, such as oils, waxes, and asphalts, can, for the most part, be ignored. Because of the refinery processes used to manufacture these products and the methods of their distribution for end use, these products can be placed into a relatively simple classification scheme based on their overall chromatographic patterns. The large numbers of compounds and their relative ratios give characteristic patterns when separated by gas chromatography that can be compared to standards and qualitatively identified.

Table 5.1. Flammable and combustible liquid classification scheme

Class number (class name)	'Peak spread' based on n-alkane carbon numbers	Examples
1 Light petroleum distillates (LPD)	C_4–C_8	Petroleum ethers, pocket lighter fuels, some rubber cement solvents, VM&P Naphtha
2 Gasoline	C_4–C_{12}	All brands and grades of automotive gasoline, including gasohols, some camping fuels
3 Medium petroleum distillates (MPD)	C_8–C_{12}	Paint thinners, mineral spirits, some charcoal starters, dry-cleaning solvents, some torch fuels
4 Kerosene	C_9–C_{16}	No. 1 Fuel Oil, Jet-A (aviation) fuel, insect sprays, some charcoal starters, some torch fuels
5 Heavy petroleum distillates (HPD)	C_{10}–C_{23}	No. 2 Fuel Oil, Diesel Fuel
0 Unclassified	Variable	Single compounds such as alcohols, acetone, or toluene. Xylenes, isoparaffinic mixtures some lamp oils, camping fuels, lacquer thinners, duplicating fluids, others

Table 5.1 shows the classification scheme adopted by the American Society of Testing and Materials (ASTM) in 1991 (ASTM E-1387-90). This classification is a refinement of the scheme originally proposed by studies of the National Bureau of Standards and the Bureau of Alcohol, Tobacco and Firearms in 1982 (*Arson Analysis Newsletter*). The scheme is based on retention time windows or peak spread of the compounds eluting as defined by *n*-alkane carbon number. Evaluation of these products requires gas chromatographic conditions that can separate compounds from butane (C4) to tricosane (C23). Figure 5.1 shows the separation of a test mixture

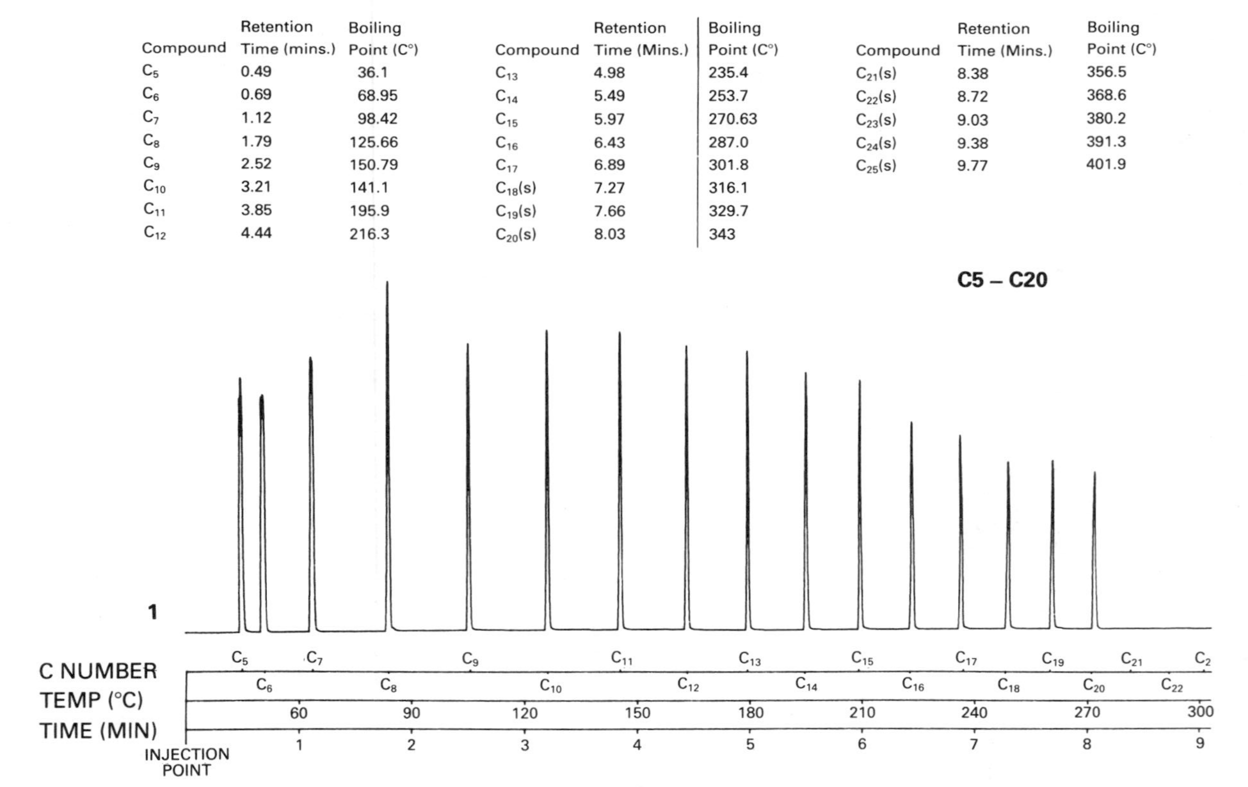

Compound	Retention Time (mins.)	Boiling Point (C°)	Compound	Retention Time (Mins.)	Boiling Point (C°)	Compound	Retention Time (Mins.)	Boiling Point (C°)
C_5	0.49	36.1	C_{13}	4.98	235.4	C_{21}(s)	8.38	356.5
C_6	0.69	68.95	C_{14}	5.49	253.7	C_{22}(s)	8.72	368.6
C_7	1.12	98.42	C_{15}	5.97	270.63	C_{23}(s)	9.03	380.2
C_8	1.79	125.66	C_{16}	6.43	287.0	C_{24}(s)	9.38	391.3
C_9	2.52	150.79	C_{17}	6.89	301.8	C_{25}(s)	9.77	401.9
C_{10}	3.21	141.1	C_{18}(s)	7.27	316.1			
C_{11}	3.85	195.9	C_{19}(s)	7.66	329.7			
C_{12}	4.44	216.3	C_{20}(s)	8.03	343			

Fig. 5.1 Chromatogram of a normal alkane standard on a 100% methyl silicone 15 m × 0.25 mm i.d. capillary column column, 1 μm coating thickness. Carrier gas, helium at approximately 3 cm^3/min. Temperature program, 60°C for 1 minute, ramp at 30°C/min to 280°C for 5 minutes. Split ratio, 20:1.

of normal alkanes that provides sufficient separation and resolution for identification of the products listed in this classification scheme.

The classification scheme divides petroleum products into five classes with an 'unclassified' class for single compounds such as alcohols, acetone, or toluene. This class also includes such specialty products as isoparaffinic (branched alkanes) mixtures that do not fall within any of the five classes. With the exception of gasoline, the identification of petroleum products in fire debris is constrained to these broader classes because of the industry's production and marketing practices. Gasoline, a blended petroleum product, produces a chromatographic pattern distinctive enough to place it in a class by itself. However, this is not the case for the majority of commercial petroleum products. Many petroleum products are blended, compounded, or otherwise modified materials from the original refinery stock. Also, there can be more than one trade name for the same basic product depending on end use. For example, the same refined product may be sold to the paint industry as mineral spirits and to the dry-cleaning industry as Stoddard solvent, provided that it meets specifications for both applications (Gibbs & Hoffman 1968). Further discussion on the use of this classification scheme will be made under the section on chromatographic interpretation.

Gas chromatography is useful for analyzing samples for products other than the more commonly encountered flammable and combustible liquids. One delay used by arsonists is a candle. The candle may be intended to ignite natural gas when a flammable fuel/air mixture is present, or could be used to ignite some other material as the candle burns down, such as material in a trailer. The identification of wax in the debris may provide information on the mode of ignition of the fire, or confirm the story of informants or witnesses. Ettling (1974) reported a method of recovering a variety of waxes from fire debris, with gas chromatographic separation on a packed column with a temperature program from 175°C to 320°C at 9°C/minute. Passing extracted samples through activated alumina prior to gas chromatographic analysis eliminated some interfering natural products. Waxes were extracted from simulated debris, using chloroform or hexane. Ettling was able to distinguish waxes from different sources. Most candle wax is made from crude oil. These waxes could be distinguished from waxes of natural sources, where a predominance of odd-numbered aliphatic hydrocarbons were present.

Capillary gas chromatography has been used to analyze and differentiate petroleum-based lubricants. Lloyd (1982) reported being able to differentiate ten different automotive engine oils based on gas chromatographic pattern differences. Using a 50 m × 0.25 mm OV1 glass column with a temperature program from 80–170°C at 15°C per minute, then to 280°C at 4°C per minute, the oils gave characteristic unresolved envelope patterns. Gas chromatographic patterns of the oils were more discriminating than fluorescence or infra-red analysis. One case example is cited where the technique was used to confirm the presence of a highly refined white mineral oil derived from a cosmetic product.

5.2.4.2.2 *Chromatographic analysis conditions*

Gas chromatography provides the ability to qualitatively identify trace amounts of petroleum products recovered from fire debris. Because of the large number of compounds present in these products, they produce characteristic chromatographic patterns that can be identified by pattern recognition techniques (Clark & Jurs 1975, 1979). Forensic laboratories initially used gas chromatographs with packed columns and thermal conductivity detectors (TCD). The thermal conductivity detector is a universal detector, meaning that it detects almost all compounds. One problem with using it for fire debris analysis is its high sensitivity to water and low relative sensitivity to hydrocarbons. Fire debris samples are often water saturated, and the presence of a large water peak often obscures the hydrocarbon peaks, preventing identification using pattern recognition. Most forensic laboratories today use a flame ionization detector (FID) which has a high sensitivity for hydrocarbons and similar compounds and has virtually no response to water (Sullivan 1977).

Packed columns were originally used for separation of samples recovered from fire debris. While interpretable chromatograms are obtained by using columns as short as 6 feet, longer columns from 12 to 20 feet provide improved peak separation with little increase in analysis time (Midkiff 1982). Packing materials coated with nonpolar phases are most commonly used for separation of the hydrocarbons. Polar phases also found applications for the separation and comparison of aromatic compounds found in gasoline (Midkiff 1980). Leung & Yip (1970) reported using a Bentone 34 column impregnated with 10% di-isodecylphthalate (DIDP) for the analysis of gasoline. Since the chromatographic features of gasoline are dominated by the aromatic constituents, this polar column gives good separation of these compounds, including the xylene isomers which are not completely separated on a nonpolar stationary phase. Yip & Clair (1976) later reported improved separation by using a 1:1 mixture of Bentone 34 and DIDP on Chromosorb. A similar separation could be done on polar capillary columns. One manufacturer has reported the separation of the aromatic compounds of gasoline, including the xylene isomers, in less than 30 minutes on a polar phase fused silica capillary column (*The Supelco Reporter* 1988). No applications of this column in fire debris analysis have been reported in the literature.

The first reports published on the use of gas chromatography for identification of petroleum products in fire debris samples used isothermal programming (Lucas 1960, Parker *et al.* 1962). With the advent of instrumentation capable of temperature programming, Midkiff & Washington (1972) reported improved resolution for products containing higher boiling compounds by using temperature programming from 50°C to 150°C. Today, most laboratories use temperature programming to effectively separate the wide boiling point range of mixtures found in petroleum products.

Capillary columns are gaining in popularity for the separation of petroleum products. Sanders & Maynard (1968) identified over 250 hydrocarbon compounds in motor gasoline, using a 200 foot stainless steel column and mass spectroscopic detection. Armstrong & Wittkower (1978) reported the use of a 100 foot capillary column in the identification of flammable liquids in fire debris. The original capillary columns introduced in 1957 were difficult to make and use, and found little use in the forensic laboratories. Fused silica capillary columns were introduced in 1979. They are flexible, convenient, and commercially available. The high resolution combined with short analysis times has made their use popular in most areas of gas chromatography (McNair *et al.* 1985). In 1983, wide-bore capillary columns were introduced. These columns have become a popular replacement for packed columns since they do not require a capillary inlet to split the sample or make-up gas for most detectors. Wide-bore columns were shown to be a useful alternative to packed columns in flammable liquid analysis, having the advantage of direct injection while providing high column efficiency (Brettell *et al.* 1986). The summary of the 1989 Flammable Liquid Proficiency Test survey administered by Collaborative Testing Services shows that 105 of 110 responding laboratories are using narrow and wide-bore capillary columns for fire debris analysis. Four laboratories reported using packed columns, while one laboratory reported using a packed column for analysis of headspace samples and a capillary column for liquid extracts. Capillary and wide-bore columns have now become the standard gas chromatography column used in petroleum product identification from fire debris samples.

The particular column and chromatographic conditions used are not as important as is the resolution needed to produce a chromatographic pattern containing sufficient points of comparison so that an identification can be made. ASTM Standard Test Method E-1387-90 recommends that a test mixture of equal parts by weight of even-numbered normal alkanes ranging from *n*-hexane through *n*-eicosane, plus toluene, *p*-xylene, *o*-ethyltoluene, *m*-ethyltoluene, and 1,2,4-trimethylbenzene need to be resolved. The main advantage of a short capillary column, for example, 15 m, is short analysis time. A sample of gasoline elutes in 5 minutes on a 15 m methylsilicone column as opposed to twenty minutes on a 20 foot packed column, with increased resolution (Fig. 5.2).

Capillary and wide-bore columns also offer more flexibility than packed columns in terms of separation requirements. Because the column efficiencies are less sensitive to carrier gas flow rate, the same column can be used for a variety of resolution requirements. For example, the sample of gasoline shown in Fig. 5.2 was separated by using a relatively high flow rate, approximately 3 cm^3/min, and a fast temperature program, 30°C/minute, used to achieve a short analysis time of five minutes. Greater resolution without loss in efficiency can be achieved with this same column by slowing down the flow rate and using a slower temperature program ramp. This may be desirable in situations where the pattern is complicated by the presence of pyrolysis products from the sample substrate and enhanced separation makes interpretation easier.

Another alternative for obtaining increased separation is the use of a longer capillary column. While increased separation is achieved at the sacrifice of time, some situations, such as interference from pyrolysis products, may benefit from attaining increased resolution. One manufacturer has shown that a 100 m column is suitable for separating xylene isomers, light hydrocarbon gases at subambient temperature, and baseline resolution of hundreds of compounds in petroleum products (*The Supelco Reporter* 1989). While analyses times of 60 to 100 minutes do not make the column practical for routine use, it may provide the increased resolution necessary for difficult to interpret chromatograms.

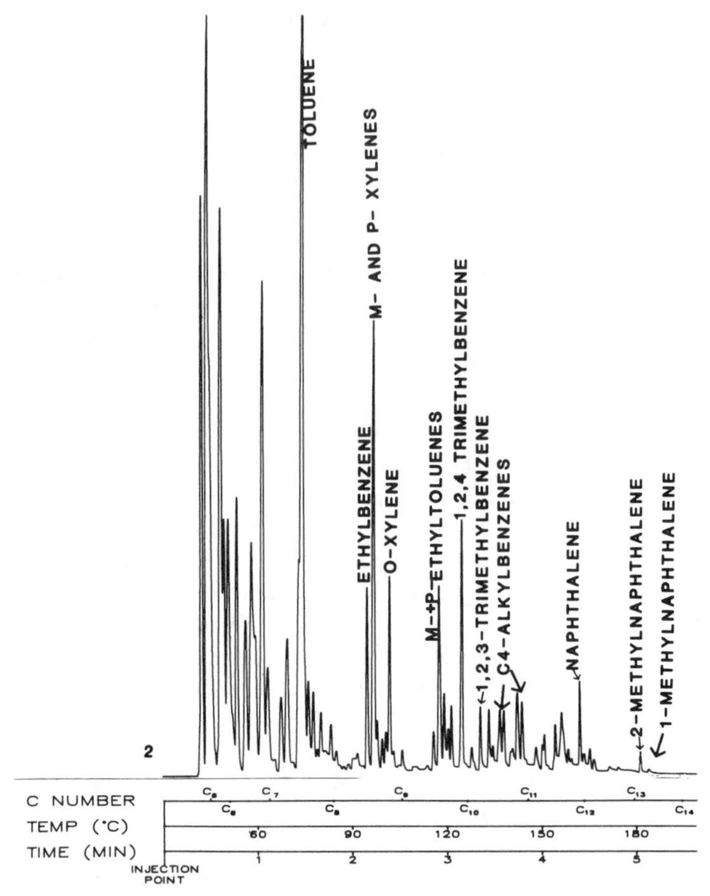

Fig. 5.2 Chromatogram of neat liquid sample of fresh gasoline separated on a 100% methyl silicone 15 m × 0.25 mm i.d. capillary column, 1 μm coating thickness. Carrier gas – helium at approximately 3 cm³/min. Temperature program – 60°C for 1 minute, ramp at 30°C/min to 280°C for 5 minutes. Split ratio – 20:1.

5.2.4.2.3 *Chromatogram interpretation*

Interpretation of the chromatograms obtained from fire debris samples is a critical step in the identification of a flammable or combustible liquid in the debris. Several factors must be considered. First, before working on any real samples, the analyst should build a library of chromatograms using the column and conditions they will be using for actual samples. Exemplars from each class identified in the classification scheme in Table 5.1 should be obtained and resulting chromatograms compiled for comparison with unknown samples. The sample preparation technique most frequently used should be considered when building the library. Neat liquid samples can be compared to unknowns isolated by most adsorption/elution techniques. However, the patterns obtained from headspace samples may differ enough from neat liquid patterns to warrant building a separate library of headspace patterns if the technique is used extensively.

Flammable or combustible liquids recovered from fire debris have normally been exposed to heat and therefore have lost some of the more volatile compounds through evaporation. The library of standards should include standards of common products in various stages of evaporation (Guinther *et al.* 1983). Recommendations for gasoline include standards of unevaporated, 20%, 50%, 75%, 90%, 98%, and 99% evaporated gasoline. In highly evaporated gasoline (over 90%) several peaks in the C_8–C_{12} ranged normally associated with normal alkanes in medium petroleum distillates, may dominate the chromatographic pattern. However, these major peaks of a highly evaporated gasoline sample are usually naphthalene and 2-methylnaphthalene, not normal alkanes. Comparison of this standard with a normal alkane standard will show that the retention times of the major peaks are different from the normal alkanes in that region. This difference in retention time along with the overall relative ratios of peaks present can confirm the presence of even highly evaporated gasoline and differentiate it from a medium petroleum distillate. Trace amounts of decane and undecane may also be noted in a highly evaporated gasoline sample. Likewise, evaporated standards of camping stove fuels, kerosene, and diesel fuels can aid in pattern recognition.

Table 5.2 gives a checklist that the analyst can use in obtaining a chromatogram suitable for pattern recognition interpretation and steps used to interpret the chromatogram. The first step involves obtaining a chromatogram in which all the major peaks are on-scale. Since pattern recognition involves not only comparison of retention times but also evaluation of relative peak-height ratios, it is important to have a chromatogram that illustrates all the relative ratios of peaks present. With the advent of computerized data collection systems, it is rarely necessary to re-run a sample to obtain a suitable chromatogram for comparison purposes.

The first step in interpretation of the chromatogram is noting in what area peaks appear. This is noted in the classification scheme in Table 5.1 as peak spread based on *n*-alkane number, which approximates boiling point of the compounds eluting from the column. Since actual retention times vary with column and gas chromatographic conditions, defining peak spread to *n*-alkane elution region provides a standard that applies to all conditions.

Light petroleum distillates appear in the C_4 to C_8 region of the chromatogram. Fig. 5.3 shows an example of a typical LPD. They can be a straight cut from the distillation tower composed mainly of aliphatic hydrocarbons, or may be specialty products resulting from blending of aromatic hydrocarbons alone or with aliphatic distillates. This class may be difficult to detect because a narrow cut provides fewer peaks and points of comparison and also, being a highly volatile fraction, is more likely to be totally consumed in a fire.

Gasoline, being a blended product of a straight-run distillate cut with aromatic compounds added to boost the octane rating, is placed in a class by itself. Fig. 5.2 illustrates the predominance of the aromatic compounds in the gasoline chromatographic pattern. Gasoline has a boiling point range of 100–400°F. The *m*-ethyltoluene/1,2,4-trimethylbenzene peak grouping is characteristic of gasoline, and can

Table 5.2. Checklist for interpretation of chromatograms

1.	Obtain a chromatogram in which the major peaks are 'on-scale' (3/4 to full scale).
2.	Note the chromatographic conditions and verify correct operation of the gas chromatograph. Run standards and blanks. a. carrier gas velocity b. temperature program c. chart speed d. attenuation e. column type and dimensions f. sample preparation technique used
3.	Mark and line up injection point on the unknown and standard chromatogram.
4.	Note in what area of the chromatogram peaks are present. a. In what retention time range are peaks present, that is light, medium, or heavy? b. What is the width of the range the peaks are in, narrow or wide?
5.	Once an area of interest is identified, look for characteristic features of a class of petroleum products. For example: a. Light petroleum distillate (C_4–C_8): – elute early – may have a narrow boiling range b. Medium petroleum distillate (C_8–C_{12}): – 2 to 3 normal alkanes usually present c. Gasoline (C_4–C_{14}): – ethyl toluenes and 1,2,4-trimethylbenzene grouping present and all peaks in correct relative ratios – C_4-alkylbenzenes present and in correct relative ratios – naphthalene and methylnaphthalenes present d. Heavy petroleum distillate (C_9–C_{25}) – normal alkane series present – pristane and phytane present in diesel (no. 2) fuel oil
Note:	Use these features to limit your possibilities. Rule out some classes of petroleum products and focus on features of remaining possibilities.
6.	When comparing questioned and standard chromatograms, line up the injection points. Match peaks in the standard to peaks in the questioned chromatogram for both retention time and relative ratios of peak height. MAKE SURE BOTH ARE RUN UNDER THE SAME CHROMATOGRAPHIC CONDITIONS!!
7.	Remember it may be necessary to re-run or replot the sample at different attenuations to look at minor peaks, or use a different program giving higher resolutions to clarify an area of interest.

be seen in even highly evaporated samples. The C_4-alkylbenzene, naphthalene, and methyl-naphthalene peaks are also important characteristic features in the gasoline chromatographic pattern. All these peaks, in the correct relative ratios, accounting

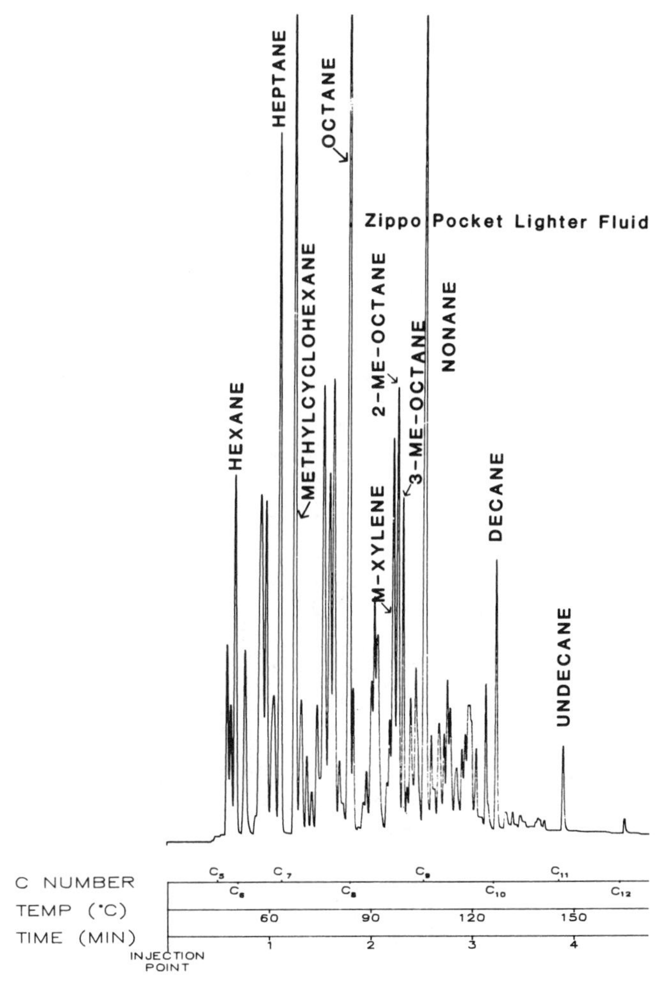

Fig. 5.3 Chromatogram of a neat liquid sample of Zippo® pocket lighter fluid, a light petroleum distillate, separated on a 100% methyl silicone 15 m × 0.25 mm i.d. capillary column, 1 μm coating thickness. Carrier gas – helium at approximately 3 cm^3/min. Temperature program – 60°C for 1 minute, ramp at 30°C/min to 280°C for 5 minutes. Split ratio – 20:1.

for normal evaporation, should be present in a pattern resulting from gasoline.

Medium petroleum distillates normally elute in the C_8 to C_{12} range and are characterized by two to three *n*-alkanes as the major components. This is illustrated in Fig. 5.4. A wide variety of commercially available products contain this class of petroleum products. These include paint thinners, mineral spirits, dry-cleaning solvents, some charcoal lighter fluids and torch fuels.

Another group of petroleum products that elute in this region is the isoparaffin solvents. These products are composed entirely of branched alkanes. They are sold by Exxon under the trade name of ISOPARS™ and were developed to have minimal

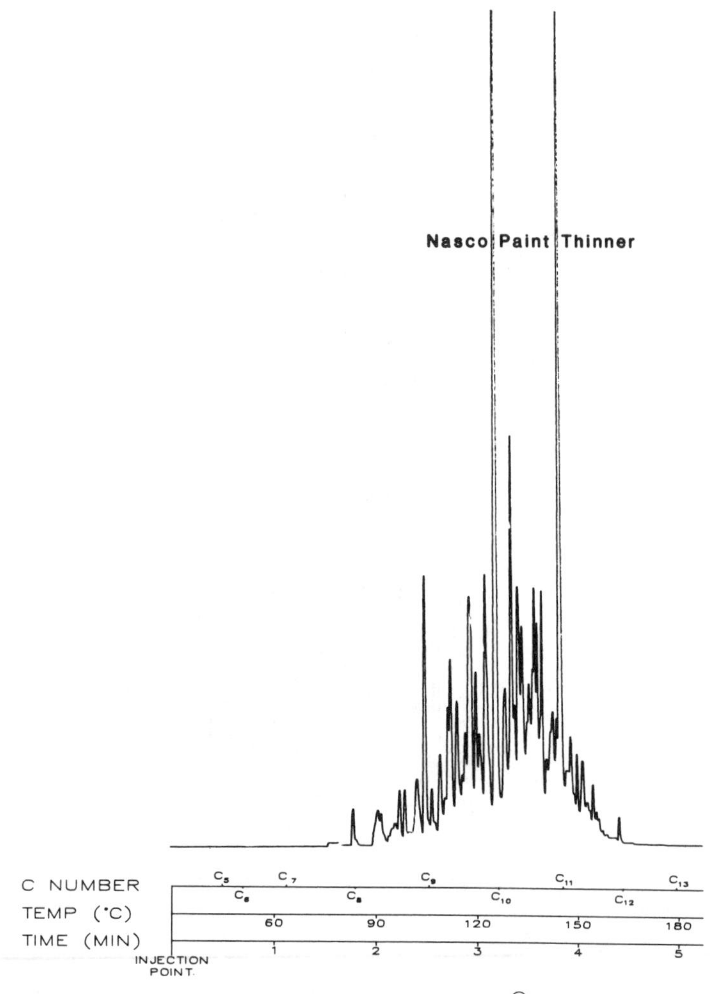

Fig. 5.4 Chromatogram of a neat liquid sample of Nasco® paint thinner, a medium petroleum distillate, separated on a 100% methyl silicone 15 m × 0.25 mm i.d. capillary column, 1 μm coating thickness. Carrier gas – helium at approximately 3 cm^3/min. Temperature program – 60°C for 1 minute, ramp at 30°C/min to 280°C for 5 minutes. Split ratio – 20:1.

or no odor, be colorless, and chemically stable. Eight different volatility grades are available, some of which elute in the medium petroleum distillate region. These products produce a chromatographic pattern that is readily distinguishable from distillation cuts (Garten 1982). Fig. 5.5 shows an example of an isoparaffinic product. Comparing this to the medium range distillate in Fig. 5.5 one can easily see the differences between the two products. The isoparaffins do not show an unresolved envelope, and the chromatogram is not dominated by two to three normal alkanes. Isoparaffinic solvents have found use in commercially available products such as electrostatic copier fluids, charcoal lighter fluids, and some lamp oils. Savin™ copier

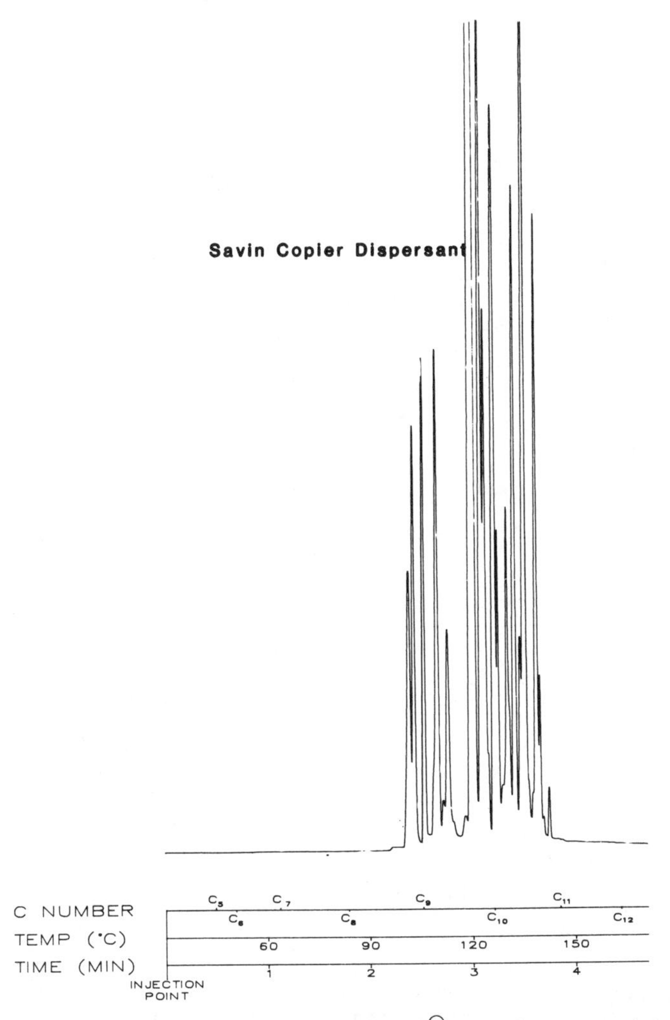

Fig. 5.5 Chromatogram of a neat liquid sample of Savin® copier dispersant, an isoparaffinic petroleum product, separated on a 100% methyl silicone 15 m × 0.25 mm i.d. capillary column, 1 μm coating thickness. Carrier gas – helium at approximately 3 cm³/min. Temperature program – 60°C for 1 minute, ramp at 30°C/min to 280°C for 5 minutes. Split ratio – 20:1.

fluid and Gulflite™ charcoal starter fluid are two examples of commercial products made from these solvents. Under the new ASTM classification scheme, these synthetically made products are placed in the Unclassified class to distinguish them from products distilled from crude oil.

Products in the fourth class, named kerosene, are distillation products that elute between C_9 and C_{16} with the homologous series of normal alkanes dominating the chromatographic pattern (Fig. 5.6). These products have a boiling point range of 377–483°F. Kerosene, diesel fuel no. 1, Jet-A aviation fuel, some charcoal lighter fluids and insecticide spray carriers cannot be distinguished from one another on the

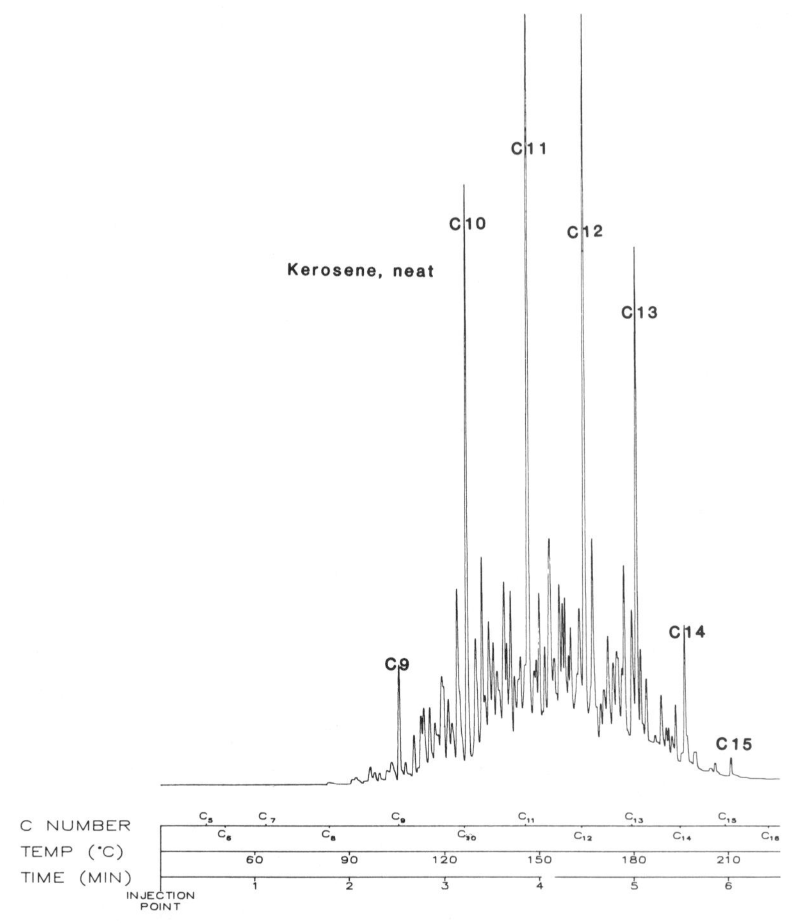

Fig. 5.6 Chromatogram of a neat liquid sample of kerosene, separated on a 100% methyl silicone 15 m × 0.25 mm i.d. capillary column, 1 μm coating thickness. Carrier gas – helium at approximately 3 cm^3/min. Temperature program – 60°C for 1 minute, ramp at 30°C/min to 280°C for 5 minutes. Split ratio – 20:1.

basis of chromatographic pattern recognition, especially when recovered from fire debris. While each of these products must meet certain ASTM specifications, they are related to physical and performance properties such as volatility, flash point, viscosity, and sulfur and water content (Dyroff, 1989). These specification differences are not reflected in the chromatographic patterns, and therefore all these products are placed in the same class.

Similarly, the Heavy Petroleum Distillates (Class 5), which includes distillates which elute from C_{10} to C_{23}, contain chromatographically similar commercial products such as diesel fuel and fuel oil no. 2 (home heating oil). They have a boiling point range of 434–595°F. The homologous series of normal alkanes are the dominating feature of the chromatographic pattern. Fig. 5.7 is a chromatogram of

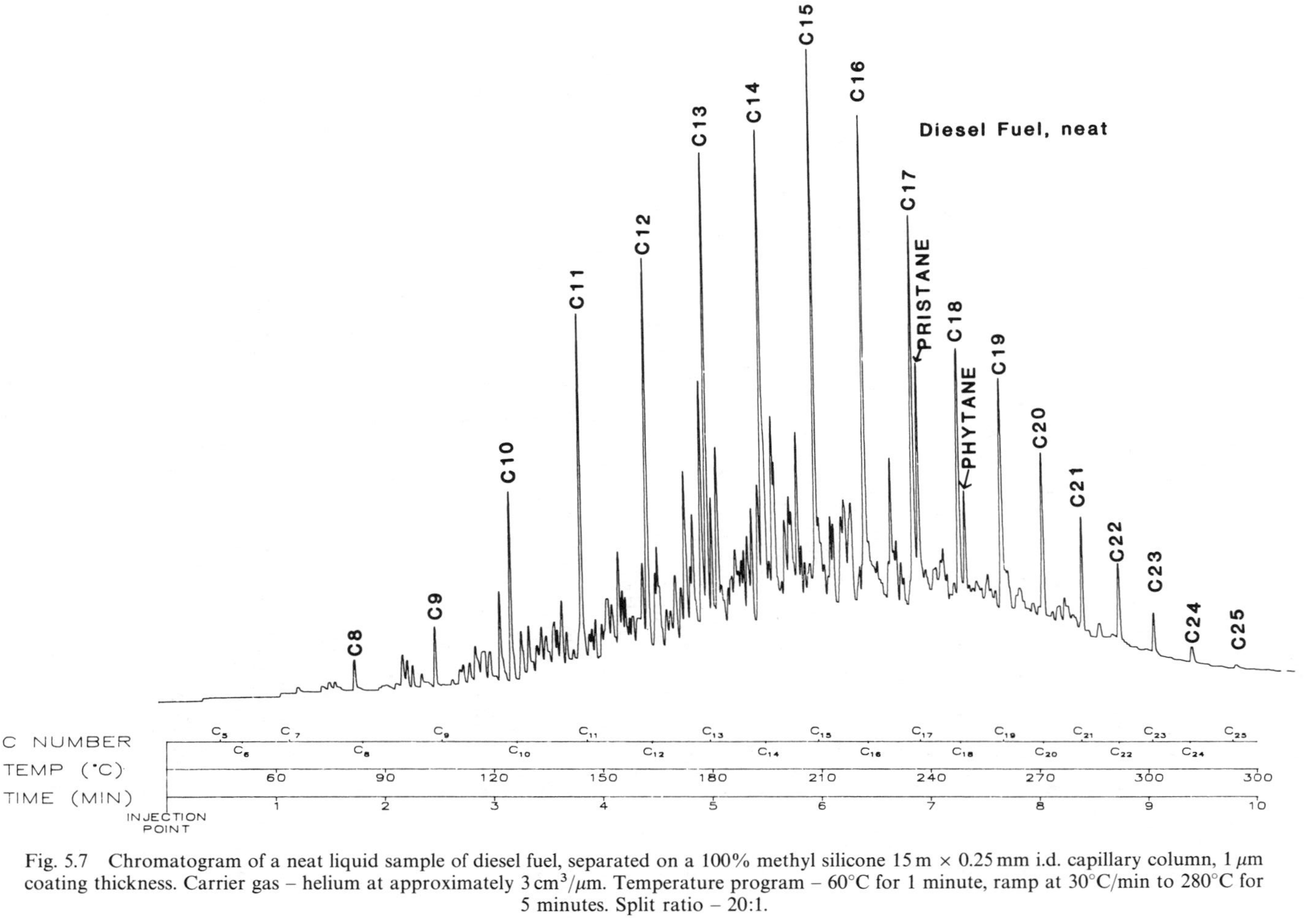

Fig. 5.7 Chromatogram of a neat liquid sample of diesel fuel, separated on a 100% methyl silicone 15 m × 0.25 mm i.d. capillary column, 1 μm coating thickness. Carrier gas – helium at approximately 3 $cm^3/\mu m$. Temperature program – 60°C for 1 minute, ramp at 30°C/min to 280°C for 5 minutes. Split ratio – 20:1.

unevaporated liquid diesel fuel. The neat liquids of this class are readily distinguishable from those in the kerosene class because of the much wider retention time window that they cover. Pristane (2,6,10,14-tetramethyl-pentadecane) and phytane (2,6,10,14-tetramethyl-hexadecane), which elute after C_{17} and C_{18} respectively, are present in all crude oil distillates and are a characteristic feature of heavy petroleum distillates and evaporated kerosene class distillates.

It may not be possible to identify products in the unclassified class by GC-FID pattern recognition. In the case of single compounds such as alcohols, acetone, and toluene, identification cannot be made on the basis of retention time on a single column alone. Other techniques such as GC/MS or GC-FTIR may be needed to confirm these compounds. However, some specialty products on the markets such as the isoparaffinic solvents contain sufficient points of comparison for identification by pattern recognition. The wide variety of commercially available products containing many different solvents emphasizes the need for each laboratory doing fire debris analysis to develop its own library of standards. A wide variety of commercial products whose labels indicate that they are flammable or combustible, or contain petroleum distillates, naphthas, or other indications of solvents, should be continuously procured and standards of their chromatographic patterns maintained.

Unknown chromatograms rarely match standards exactly peak for peak. Samples obtained from fire debris will most likely show some loss of the volatile compounds, often referred to as 'weathering'. Also, depending on the sample preparation technique used to isolate the sample, the full range of a combustible or flammable liquid may not be recovered. For example, all headspace techniques, both static and dynamic, usually do not recover hydrocarbons beyond the C_{17} range. The important point to remember is that compounds should not be selectively missing. In other words, one can explain the loss of more volatile compounds through evaporation. However, if more volatile compounds are present, but several less volatile compounds are missing while some are present, the analyst should immediately be suspicious that the pattern may be the result of volatile compounds from a source other than a flammable or combustible liquid. Even in evaporated samples, all the peaks in their known relative ratios should be present before an identification can be made.

Occasionally, it would be useful in an investigation to determine if the residue of gasoline detected at a fire scene was evaporated as a result of the original product being burned or the result of normal evaporation at moderate temperatures. A study conducted to answer this question has been reported (Mann 1990). The chromatographic patterns of gasoline evaporated at moderate temperatures and subject to combustion were compared quantitatively. Evaporated and combusted gasoline residues could not be differentiated on the basis of their chromatographic patterns. Therefore, no conclusion can be made about the cause of loss of volatile compounds in gasoline residues.

One exception to this rule of thumb has recently been reported (Mann & Gresham 1990). Flammable liquids recovered from soil samples were found to lack some of the diagnostic features associated with their chromatographic patterns. These missing features could not be attributed to chemical or physical processes. A review of the literature showed that crude oil is degraded in the environment by bacteria. Mann

& Gresham were able to demonstrate that bacteria in soil show a preferential consumption of *n*-alkanes and substituted benzenes with least substitution. The chromatographic patterns recovered from unsterile soil samples stored more than six days showed distinct differences from the original liquid.

Chromatograms may be further complicated by the addition of compounds from the debris itself or pyrolysis and combustion products in the debris. Pyrolysis can be defined as the transformation of a substance into one or more other substances by heat, whereas combustion involves a reaction between oxygen and fuel (Lentini *et al.* 1989). Fire debris samples often contain volatile combustion and pyrolysis products that are detected along with any petroleum products that may be present. When these products elute in the same retention time windows as petroleum products, interpretation of the chromatogram is more difficult. Indeed, a large amount of pyrolysis products can effectively obscure or mask the chromatographic pattern of any petroleum product present in the sample. The chromatographic pattern resulting from burned carpet and padding, illustrated in Fig. 5.8, was isolated by using passive adsorption on charcoal, and illustrates the complicated patterns that can be obtained from modern home furnishing substrates. Clodfelter & Hueske (1977) reported recovering flammable liquids on steam distillation of fire debris. However, chromatographic analysis of the liquids did not give a pattern that could be confused with common petroleum products. Similarly, DeHaan & Bonarius (1988) chromatographically studied samples isolated by both heated headspace and charcoal trapping from burned floor-covering, carpet, synthetic turf, vinyl floor covering and underlay. They concluded that most of the pyrolysis products studied could be distinguished from those of common petroleum products and most synthetic blends such as lacquer thinners and enamel reducers.

Pyrolysis products from polyethylene and some other plastics may produce chromatographic patterns that could be mistaken for kerosene or heavy petroleum distillates to the untrained eye. These polymeric products break down on heating, and may produce a chromatographic pattern showing a homologous series of peaks (Iglauer 1972, Irwin 1985). Close inspection of the chromatogram reveals that many of the homologous series peaks are doublets and/or triplets. Figure 5.9 shows the chromatographic pattern resulting from the pyrolysis of polyethylene. The dominant peak of each cluster does not match the retention time of normal alkanes in the petroleum distillate, but is slightly shifted. This is because the homologous series of peaks from polyethylene are not normal alkanes, but are the geminal alkyl-diene, 1-alkene, and the *n*-alkane, eluting in that order. Polyethylene pyrolysis product chromatograms will not show the characteristic pristane and phytane peaks seen in distillate chromatograms. If the analyst has a chromatogram showing homologous series of doublets and triplets and the absence of pristane and phytane, the presence of plastic in the debris sample can explain the observed chromatographic pattern.

Many fire debris samples contain wood. Terpenes naturally present in wood are normally recovered from these samples, and can give a chromatographic pattern similar to commercially available turpentine products. The observation of such a pattern recovered from a sample containing wood presents a dilemma for the analyst; is it naturally occurring or a manufactured product added to aid the spread of the

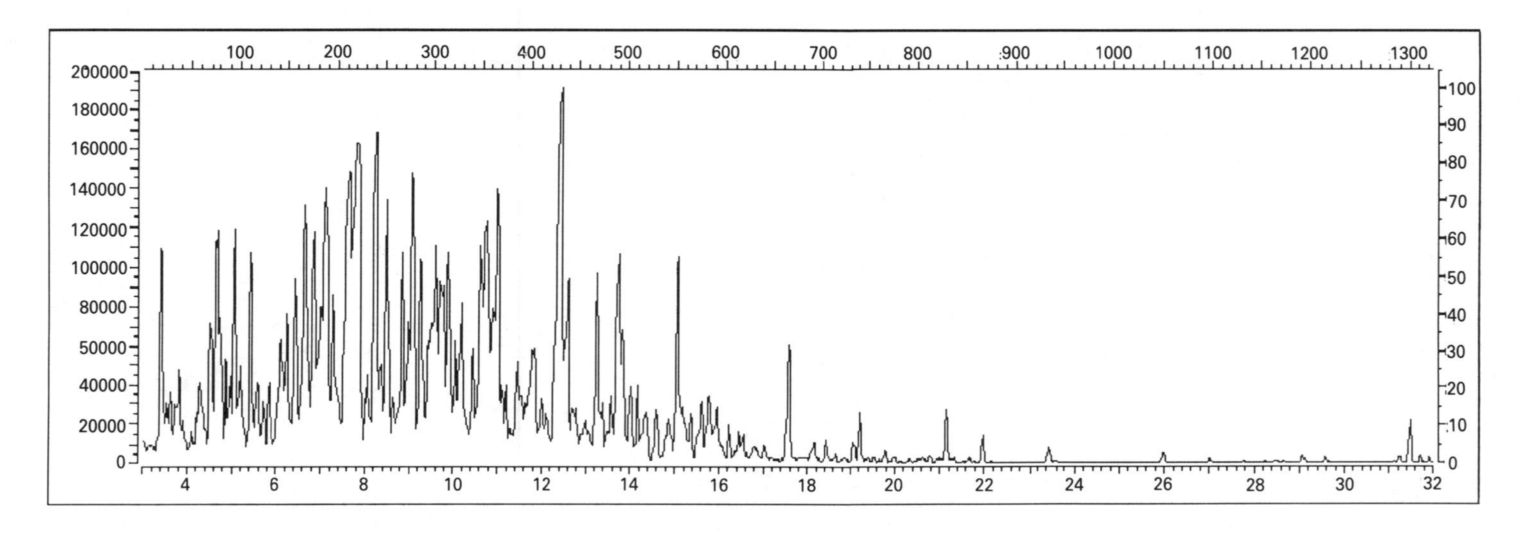

Fig. 5.8 Total ion chromatogram of sample recovered from charred carpet and padding by passive charcoal adsorption

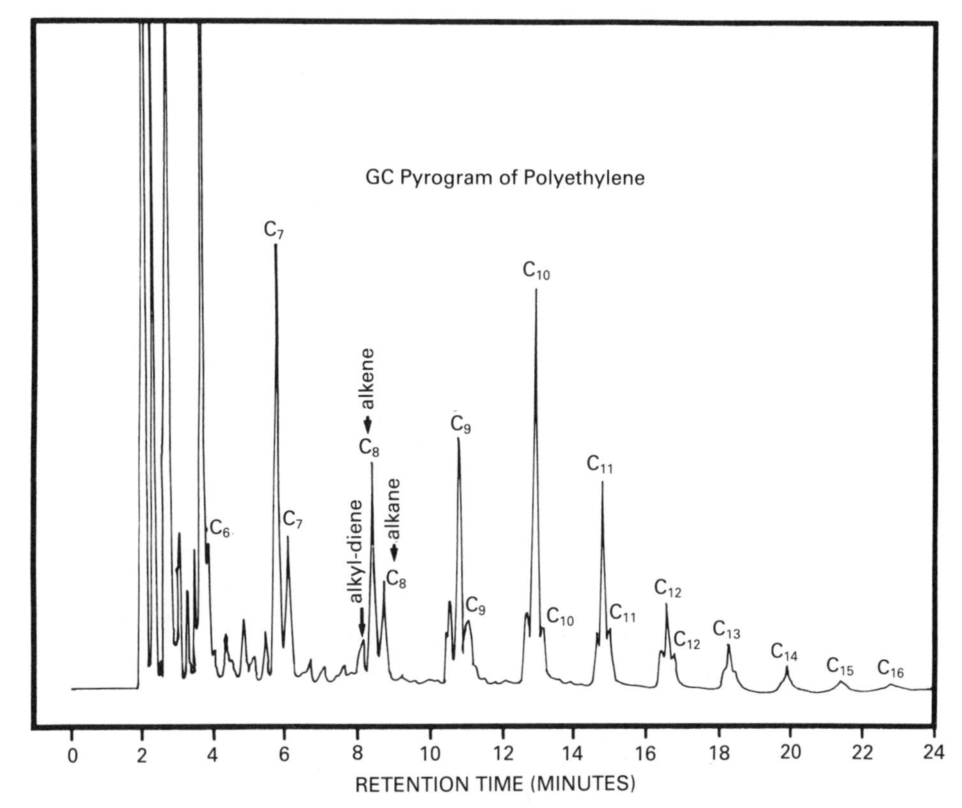

Fig. 5.9 Chromatogram of pyrolysis products from polyethylene.

fire? Trimpe (1991) performed a study on terpenes found in wood samples. He found that chromatographic patterns resembling turpentine were recovered from samples containing soft woods, mostly yellow pine. These patterns could not be distinguished from commercially available products. While hard woods contained terpenes, they were not the same compounds and produced a different pattern than commercially available turpentine products.

Interference in interpretation of chromatograms by the presence of pyrolysis products is a reality that must be dealt with routinely. Furnishings composed of synthetics often produce many volatile compounds on burning. Several tactics may be used to distinguish between pyrolysis products and flammable liquid patterns. A more selective detector, such as GC/MS, can be used to discriminate between compounds of flammable liquid origin and pyrolysis products. Care must be taken in interpreting mass spectral data since pyrolysis products may be hydrocarbons. As mentioned previously, longer columns that provide increased resolution or columns with a different stationary phase to alter elution order can also be used to increase discrimination.

Some analysts find the use of comparison samples (formerly commonly referred to as control samples) helpful in determining what peaks in the chromatogram can be attributed to substrate origin as compared to those which may be of a flammable liquid origin (Nowicki 1981). If fire investigators have taken samples from an area where they believe a liquid may have been poured, they also have been encouraged to submit samples of similar substrate taken from an area away from the suspected pour pattern. By comparing the patterns obtained from the comparison sample to that of the suspect samples, it is sometimes possible to identify peaks in the unknown that can be attributed to pyrolysis products of the substrate material. Several problems can be encountered with comparison samples. First, samples obtained away from a pour pattern may actually contain unburned flammable liquid that was poured but did not ignite. Second, the pattern of pyrolysis products obtained from substrates is dependent on the environmental conditions, that is, amount of heat and length of burning, to which it is exposed. For example, the same piece of carpet, burned under different conditions, can give different pyrolysis peak patterns.

The Forensic Science Committee of the International Association of Arson Investigators recently published a position on the use and value of comparison samples (Lentini *et al.* 1990). First, the Committee's position states that the use of the word 'control' is not appropriate to a fire scene. The only way to obtain a true control sample is to sample the scene before the fire has been set or occurred. Therefore the use of 'comparison sample' more accurately reflects the purpose of the sample, to help the analyst understand the composition of materials inherent to the fire scene. Second, a comparison sample is sometimes but not always necessary for the analyst to properly interpret patterns isolated from fire debris. Obviously, the field investigator collects the samples, so they should collect comparison samples whenever it is feasible. The Committee also recognizes that it is not always possible to collect a proper comparison sample.

The brand identification of gasoline or the determination of common source between known and unknown samples is of interest to fire investigators. If the forensic analyst could report a particular brand and octane rating, this information could provide investigative leads. Similarly, if samples from the scene, either neat liquids or vapors, could be compared to gasoline obtained from a suspect, and a common source identified, a strong circumstantial link between suspect and crime scene could be established. Unfortunately, brand identification is generally considered to be impossible and determination of common source is extremely difficult (Midkiff 1975). The main reasons are the marketing practices of the petroleum refinery industry and changes which occur in the composition of products during storage. Most oil companies have exchange agreements that allow Brand A and B's trucks to fill up at Brand C's terminal. Since proprietary brand additive packages are added only at the terminal as the gasoline is being pumped into the delivery trucks, the actual refinery source of the base gasoline stock is generally not of importance to the distributing company.

Several points of comparison between neat liquid samples can be made. When both leaded and unleaded gasolines were widely used, the absence or presence of organolead compounds could be made. Gas chromatography/mass spectrometry is useful for determining the presence of organolead compounds (Kelly & Martz 1984). Tetraethyl lead elutes near naphthalene in gasoline. This organolead compound can be identified by monitoring the fragment ions, amu 208, 237, 266, and molecular ion, amu 295. With the decreased marketing of leaded gasolines, most gasoline samples analyzed in the forensic laboratory today are unleaded. Sample preparation method must be considered when looking for organolead compounds recovered from debris samples. Charcoal irreversibly adsorbs organolead compounds, therefore any samples recovered by charcoal adsorption will not contain organolead compounds (Fultz & Wineman 1988). The addition of alcohols, ketones, methyl-*tert*-butyl ether, and other oxygenates to gasoline may provide other points of discrimination between samples (See section 5.2.4.1 on specific detection). Dyes are sometimes added to petroleum products to differentiate pipeline products or as a brand signature. Extraction of the dyes from gasoline and comparison by thin-layer chromatography may give another point of comparison between samples (Pearce 1977, Long 1978; Moss *et al.* 1982).

Variation in the exact quantities of hydrocarbons present in gasoline varies for several reasons. The exact ratio of hydrocarbons is also dependent on the original crude oil from which the gasoline is refined. Higher octane gasolines contain a higher concentration of aromatic compounds. Mann (1987) reported a comparison method based on these quantitative differences that could be used to differentiate unevaporated liquid gasoline samples. Compounds eluting in the *n*-butane to *n*-octane regions have more batch to batch variations because of a greater sensitivity of these compounds to subtle changes in the refining parameters and blending processes. Chromatograms of liquid samples are obtained on a capillary column under conditions giving baseline resolution. Samples are first compared qualitatively by overlaying the chromatograms. Large to moderate differences in peak heights are readily visible and the sample is considered to have come from different sources. A quantitative peak ratioing approach is applied to samples which cannot be visually differentiated from one another. Even when using this quantitative approach, Mann concluded that the technique was powerful for stating that two gasolines did not have a common source, but the strongest conclusion that could be drawn for similar samples was that they could have had a common origin. Unless the analyst is testing within a totally known population, that is, has access to every possible source of the gasoline, the possibility of a sample coming from a gas station outside the tested population precludes a stronger conclusion.

Mann further tested this quantitative approach on samples recovered from debris (Mann 1987). A sample recovery technique which did not quantitatively change the profile of gasoline was necessary. Charcoal preferentially adsorbs and desorbs different classes of hydrocarbons, thus eluents from charcoal adsorption have slightly altered chromatographic profiles from the original liquid. Also, the area most significant for comparison between samples is lost in samples evaporated more than 40 to 60 per cent by weight. Applying a more rigorous quantitation method than that used for neat liquid samples, Mann was able to successfully discriminate between samples not having a common source. Again, as with liquid samples, the conclusive determination of common origin was not possible.

Another method used to reduce the interference of pyrolysis and combustion products is to remove them prior to gas chromatographic analyses. Several pre-analysis clean-up techniques have been reported. The separation of pyrolysis products from petroleum-based compounds depends on the pyrolysis products being chemically different species. As noted earlier, pyrolysis products themselves may be hydrocarbons. No pre-analysis clean-up technique can separate hydrocarbon pyrolysis products from petroleum derived hydrocarbons. However, some pyrolysis products are chemically different species that are also detected by the flame ionization detector. Isothermal pyrolysis of cellulose produces furan, acetaldehyde, acetal, furfural, propanoic acid, and cyclohexanone (Lipska & Wodley 1969). If these species can be removed prior to analysis, the resultant gas chromatographic pattern may be easier to interpret.

Aldridge (1981) reported a method that separated terpenes from gasoline by thin-layer chromatography. The presence of large amounts of terpenes from wood can interfere with chromatogram interpretation since they elute in the same boiling point region as the more characteristic features of gasoline. Liquid extracts are applied to a silica gel thin-layer plate and eluted with 95:5 hexane:benzene mixture. The aromatic compounds in gasoline are separated from the terpenes. The position of the aromatic compounds on the plate is visualized under ultra-violet radiation, using a standard gasoline samples. The portion of the plate containing the aromatics is scraped off, eluted with carbon disulfide, and analyzed by using routine GC-FID conditions. Aldridge & Oates (1984) later reported using silica solid-phase extraction columns to clean-up liquid extracts from fire debris. The use of solid phase extraction tubes provides simpler sample manipulation than thin-layer chromatography. Polar compounds from burning carpet material were removed, being adsorbed on the polar silica gel, while the nonpolar hydrocarbons pass through the column. The column eluent is chromatographed under normal GC conditions.

Another method which can be used to clean-up samples prior to GC analysis is acid-stripping. This procedure was originally developed by Juhala (1979) to remove oxygenated species in lacquer thinners as a way to confirm their presence in GC-FID patterns. Carbon disulfide extracts from fire debris samples are extracted with a mixture of concentrated sulfuric and phosphoric acids. Compounds containing oxygen will be extracted into the acid liquid phase. ASTM Standard Practice E-1389-90 provides a practical guide for the use of acid stripping for pre-GC analysis.

5.2.3 Sample preparation techniques

In order to identify the presence of a petroleum product in fire debris samples, some method must be used to recover the product from the sample and introduce it into the gas chromatograph. Sample preparation techniques must recover products in detectable and therefore identifiable quantities. Techniques that are efficient, simple, and rapid are desirable. Each of the sample preparation techniques described below has advantages, disadvantages, and limitations. The analyst must be aware of the effects of the sample preparation technique used on the final chromatograph obtained to properly interpret the results. A recent review of sample preparation methods used for detection of petroleum products reported that improved isolation techniques resulted in improved sensitivity and specificity of product identification (Caddy *et al.* 1991). Charcoal adsorption is now the most widely used method of isolation, because it most closely meets the features of an ideal recovery system: reasonable recovery;

separation from pyrolysis products; total sample analyzed; repeat analyses possible; preparations can be cleaned prior to analysis; sensitive; rapid and cheap.

This section briefly describes the major sample preparation techniques in use, and will summarize with a comparison of the various techniques as to advantages and disadvantages. No single method can be universally applied to all types of petroleum products and samples encountered (Bertsch & Zhang 1990). A proposed analysis scheme will show how these techniques fit into an overall analysis scheme for the identification of flammable and combustible liquids.

Liquid samples require virtually no sample preparation. They can be injected neat into the gas chromatograph. Syringes delivering as little as 5 nl are available for injecting neat samples onto capillary columns which have a small sample capacity. Alternatively, the liquid product can be diluted in an appropriate solvent prior to analysis.

5.2.3.1 Distillation

Among the first techniques used to isolate petroleum products from debris samples were distillation methods. These included simple distillation, steam distillation, and vacuum distillation (Brackett 1955). Steam distillation techniques have been modified by the addition of alcohol (Macoun 1952) and ethylene glycol (Brackett 1955). Simple distillation was found to be suitable only when low boiling products were present. Yip & Clair (1976) concluded that steam distillation was useful for recovering petroleum products if previous headspace sampling gave negative results. In a comparison of various distillation techniques, Woychesin & DeHaan (1978) found steam distillation to be the most effective for the least effort. Ethylene glycol was useful for higher boiling products, but also carried over more contamination from the substrate than other techniques. Ethylene glycol and vacuum distillations offered the highest recovery of petroleum products of the various distillation methods. Hrynchuk *et al.* (1977) reported using vacuum distillation to recover petroleum products from debris. Levels of 50 ppm of petroleum product by weight of charred debris were routinely recovered. One cited advantage of the preparation method was that the exhibit material was essentially unchanged. Not altering the material can be of importance when examining documents.

ASTM E-1385-90 Standard Practice for *Separation and concentration of flammable or combustible liquid residues from fire debris samples by steam distillation* outlines current procedures for separating visible quantities of water insoluble hydrocarbons from fire debris samples. The Standard recommends that the technique be used only for samples which have a detectable odor of petroleum distillates at room temperature. One advantage cited for the technique is that the distilled liquid can be used as a courtroom exhibit, the odor being readily recognizable by a jury.

Samples are placed in a container and an appropriate amount of water is added to the sample and brought to the boil. The vapors produced are condensed in a volatile oil distillation apparatus. Petroleum products float on top of a column of water and the visible liquid is recovered for chromatographic analysis.

5.2.3.2 Static headspace

Liquids exert vapor pressures that are dependent on their chemical composition. In a closed system, a liquid is in equilibrium with its vapor. Therefore any flammable

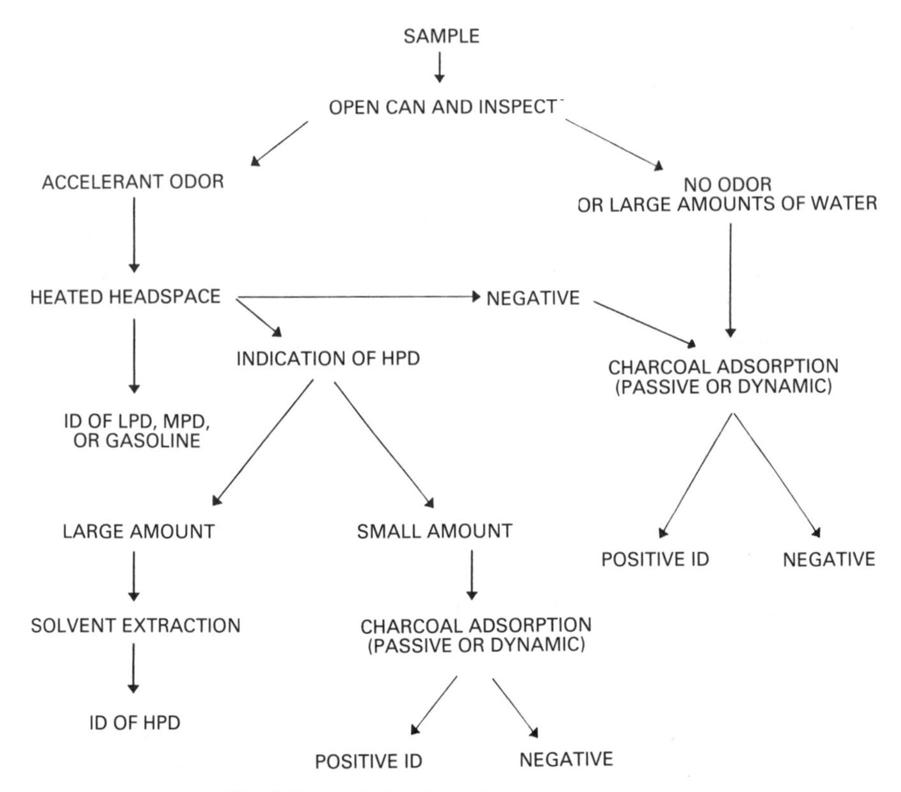

Fig. 5.10 Analysis scheme for fire debris samples.

or combustible liquid present in fire debris will vaporize, and be present in the headspace of the sample container. A small amount of this headspace can be sampled and injected into a gas chromatograph to detect the presence of such a product. Heating the sample in its original container will increase the vapor pressure and thus the concentration of vapors from the liquid in the headspace of the samples. Headspace sampling has been a method of choice for fire debris analysis because of its simplicity and speed of analysis (Loscalzo *et al.* 1977).

ASTM Standard Practice E-1386-90 for *Sampling of headspace vapors from fire debris samples* outlines the procedure for removing small quantities of flammable or combustible liquid vapors from the headspace of debris containers. The practice calls for heating the sample, preferably in its original container, prior to sampling and analysis by gas chromatography. The technique is useful for screening samples for the presence of a particular type of petroleum product, and is also useful when volatile oxygenated products such as alcohols or lacquer thinners are suspected. The practice is one of the least sensitive sample preparation methods, however, and may not detect quantities of less than 5 or 10 μl of a petroleum product in an empty gallon can (Brettell *et al.* 1984).

Some analysts use headspace sampling to screen samples at room temperature.

Others heat the samples to temperatures from 50°C to 120°C to increase the vapor pressure and thus the concentration of vapors from liquids in the debris into the headspace (Yip & Clair 1976). Samples are typically heated in an oven or on a hot plate from 10 to 30 minutes. Heating wet samples above 100°C may cause venting or rupturing of the container and loss of the sample. Another danger from overheating is charring the contents and producing additional pyrolysis products. Typically a small hole is punched in the lid of the container prior to heating and covered with tape or plugged with a rubber septum. Using a vapor tight syringe, a small sample, 0.5 to 5 cm^3 of vapor is taken from the container and injected directly into the gas chromatograph. The amount of vapor injected depends on the type of column. Capillary columns, even when using high split ratios, do not give efficient separations when more than 1 cm^3 of vapor is injected. Vapor samples up to 5 cm^3 can be injected on megabore and packed columns, which have a higher sample capacity.

Chromatograms of vapors of flammable and combustible liquids can differ considerably from liquid sample chromatograms. The most volatile compounds of the liquid residue will predominate in the headspace. The headspace pattern of a fresh sample of gasoline will be dominated by the most volatile components, and little of the characteristic heavier components will be observed. Analysts must keep these differences in mind when interpreting headspace chromatograms.

Reeve *et al.* (1986) reported a headspace sampling method using cryogenic focusing which avoids the quantitative and qualitative problems of normal static headspace sampling. After heating in an oven at 60°C for one hour, 0.5 ml of headspace vapors was injected through an on-column injector. Approximately 25 cm of the head of the column was immersed in liquid nitrogen. After the sample 'plug' was cryogenically focused at the head of the column, a process which takes approximately ten seconds, the dewar containing the liquid nitrogen was removed from the gas chromatograph oven, and the normal program commenced. This procedure gave comparable recoveries for gasoline with passive charcoal wire adsorption. Water did not interfere with the recovery or analysis. The procedure was not as sensitive for detecting diesel fuel as charcoal adsorption, as would be expected for less volatile petroleum products.

5.2.3.3 Dynamic headspace adsorption

While headspace techniques are simple and rapid, they lack in sensitivity. Distillation recovery techniques are time consuming and cumbersome. In searching for a rapid and sensitive sample preparation method, forensic scientists borrowed from the industrial hygiene industry and began using vapor concentration techniques using sorption materials.

Chrostowski & Holmes (1979) reported a procedure for the collection and concentration of vapors from fire debris samples on activated charcoal. The sampling procedure was an adaptation originally used to collect explosive vapors from explosive debris. Approximately 3 inches of 50–200 mesh activated charcoal was packed in a $5\frac{3}{4}$ inch disposable pipet. Preheated nitrogen was passed over the fire debris sample at a rate slightly greater than 3 l/minute by pull from a vacuum pump. Sample times varied from 30 to 60 minutes. The charcoal tube was eluted with 2 to 3 ml of carbon disulfide, and a 5 μl aliquot injected into the gas chromatograph for analysis. Water vapor did not interfere with the adsorption of gasoline vapors. Gasoline, kerosene, and diesel fuel samples recovered by this method were identifiable by pattern recognition.

A simpler procedure also using activated charcoal as the adsorption medium was reported (Tontarski & Strobel 1982). The amount of 50–200 mesh activated charcoal used is approximately 2.5 inches (0.5 g). Metal paint cans containing debris were punched with holes near the bottom and placed in suitable size heating mantles. The original lid was replaced with a lid containing fittings suitable for insertion of a thermometer and sampling tube containing activated charcoal. Heated room air was flushed over the debris being pulled by a vacuum on the charcoal tube. After eluting the charcoal tubes with carbon disulfide, the liquid aliquots were placed in an automatic sampler for injection into the gas chromatograph and data were collected on a computerized data system. One disadvantage of this configuration is the possibility of contamination from room air or the heating mantles. Crandall & Pennie (1984) reported a similar collection system using activated charcoal in which a tube containing charcoal is used to filter the air entering the can, thus reducing the chance of contamination from room air.

Samples collected on sorbent material can be desorbed thermally or by solvent extraction. Frenkel *et al.* (1984) looked at the relative sensitivity of headspace sampling, and sorption on charcoal with both thermal desorption and solvent washing. They reported that sorption of gasoline on the charcoal was orders of magnitude more sensitive than headspace sampling. Thermal desorption of the samples from charcoal was not satisfactory because of the high temperatures needed, 950°C, to remove hydrocarbons from the charcoal. Thermal degradation of some of the hydrocarbons is seen, thus complicating the resulting chromatographic pattern. These workers concluded that solvent extraction of the charcoal with carbon disulfide was a better method to recover petroleum products.

Juhala & Beever (1986) reported a charcoal adsorption method using disposable hypodermic needles with charcoal packed in the hub. The needle is inserted into the sample container which has been previously heated to 90°C. The container is evacuated through the needle for several minutes. Approximately 100 μl of carbon disulfide was used to elute the charcoal. Improved sensitivity using smaller amounts of charcoal and carbon disulfide were reported for both gasoline and fuel oil.

Other sorption media have been used to collect the vapors from fire debris samples. Baldwin (1977) reported using Florisil to concentrate vapors from samples, eluting the recovered compounds with carbon disulfide. One problem encountered was the loss of lower boiling compounds in gasoline. However, the chromatographic patterns obtained could be identified by comparison to liquid standards.

Tenax-GC™, a porous polymer capable of adsorbing a wide range of volatile materials, has been used to concentrate vapors from fire debris samples (Russell 1981, Saferstein & Park 1982). Tenax-GC™ had been used successfully for trace analysis of volatiles in air pollution studies, and thus appeared to be suitable for recovery of petroleum products. Russell (1981) used a syringe to pull vapors from a nylon bag containing debris over Tenax-GC™ packed in a glass tube. This tube was then placed in the injection port of the gas chromatograph for thermal desorption. Saferstein &

Park (1982) used stainless steel cartridges packed with Tenax-GC™ and flushed these cartridges with heated inert gas passed over the sample. These samples were thermally desorbed into the gas chromatograph. Water did not interfere with the sorption properties of Tenax-GC™. The method showed increased sensitivity over headspace methods of several orders of magnitude. In addition, the method is less time consuming and labor intensive than distillation techniques.

Willson (1984) reported using an automated system for headspace samples collected on Tenax. Up to 50 tubes were loaded on an automatic sampler turntable, and the tubes thermally desorbed; the resulting chromatograms were captured on a computerized data system. Headspace samples were collected by flushing warm nitrogen over debris in nylon bags and through the Tenax tubes. Automation of the gas chromatographic injection allows more samples to be run with less intervention by the analyst. This system is in use in a number of major forensic laboratories in the United Kingdom, United States, and Canada where a high throughput of samples justifies the extra costs. Like other headspace techniques, its utility is limited to LPDs and MPDs, and it is not as sensitive to HPDs.

In addition to recovery in the laboratory, Saferstein & Park used Tenax-GC™ to recover airborne petroleum products at fire scenes. They successfully recovered charcoal lighter fluid, gasoline, and kerosene at scenes up to 15 hours after extinguishment of the fire. Comparison samples of air from the scene were taken at least 30 feet away from the structure under investigation. No gasoline vapors from vehicles operating in the areas were noted.

Andrasko (1983) reported a study in which two porous polymers, Porapak Q™ and Tenax-GC™, and one chromatographic separation medium, Chromosorb 102™, were compared for effectiveness in concentrating hydrocarbon vapors. Samples were collected by passing air over heated samples in original containers, using a vacuum to pull the air through the collection tubes. Andrasko found both Poropak Q and Chromosorb 102 to be stronger adsorbents for hydrocarbons than Tenax-GC™. However, this adsorption advantage turns to disadvantage when thermally desorbing the materials for gas chromatographic analysis. Tenax-GC™ is more thermally stable than Poropak-Q™, and hydrocarbons are desorbed from Tenax-GC™ at a faster rate on heating. Tenax-GC™ is also more stable against oxidation than Porapak Q™, and more suitable for use when large amounts of sorption materials are needed. Slower desorption with Porapak Q also reduces resolution for the more volatile components introduced into the gas chromatographic. All three sorption materials gave chromatographic patterns of recovered petroleum products that could be identified by comparing them to liquid standards.

Kobus *et al.* (1987) resolved the problem of poor resolution from thermal desorption from Porapak Q by using cryogenic focusing to introduce a neat sample plug onto the chromatographic column. Adsorption onto the Porapak Q was achieved by both dynamic and passive sampling. A marked improvement in the chromatography was achieved by using cryogenic focusing.

Thermal desorption from a sorption medium is theoretically the most sensitive method since the entire collected sample is analyzed, rather than a portion obtained by liquid desorption. Polymer sorption media are not suitable for solvent desorption, since appropriate solvents would also dissolve the polymer sorption material. Thermal desorption methods have not found as widespread acceptance as sorption on charcoal. One reason may be the need for electronic data collection since the entire sample is consumed on analysis. The only way to replot the data for chromatographic interpretation is to have electronically recorded it for replotting. Also the technique is more cumbersome than solvent elution and liquid injection into the gas chromatograph.

ASTM E: 1412–91, Standard *Practice for separation and concentration of flammable or combustible liquid residues from fire debris samples by dynamic headspace concentration*, offers the analyst practical guidelines for using this sample preparation technique.

5.2.3.4 Passive headspace adsorption

The previous sorption techniques required sweeping the headspace by either positive or negative flow over the debris prior to collection on the sorption media. An alternative method of collection of vapors onto sorption media is passive sampling, that is, placing the sorption media in contact with the headspace of the sample, and simply waiting some time before removing the sorption media and desorbing compounds for analysis. This type of sample preparation is attractive because it involves very little time and labor.

Twibell & Home (1977) demonstrated that activated charcoal in contact with the headspace of debris for several hours adsorbed sufficient vapors for pyrolysis/gas chromatographic analysis. Curie pyrolysis wires were coated with finely divided activated charcoal. Samples were desorbed as usual in the pyrolysis unit. This method was found to be more sensitive than either static headspace sampling or steam distillation. In 1981, Twibell *et al.* (1981) reported a modification of the Curie point pyrolysis inlet system that could be used with capillary columns. Less charcoal needed to be used to accommodate the lower sample capacity of the capillary columns.

Andrasko (1983) studied two types of charcoal used to concentrate vapors on the Curie point wire. Finely divided charcoal powder adsorbed less volatile compounds more efficiently than granular charcoal. Conversely, granular charcoal was more efficient than powdered for adsorption of more volatile compounds. On the basis of these findings, Andrasko coated wires with both granular and powdered charcoal with sampling temperatures between 100° and 110°C and times from two to three hours giving the best results for most commonly encountered petroleum products. Mechanical stability of the wires was a problem. The wires were not robust, and some care must be taken in not destroying them when inserting them into the pyrolysis unit.

Other workers have reported using charcoal coated wires with solvent wash for introduction of a liquid sample into the gas chromatograph. Reeve *et al.* (1986) used wires made from twenty-gauge copper wire dipped in sodium silicate solution and rolled in activated charcoal. These wires were placed in sealed cans which are heated in an oven at 60°C for one hour. After heating, the wire was placed in a 1-ml test tube, and washed with 150 μl of carbon disulfide. Tranthim-Fryer (1990) reported improved results when using a charcoal coated wire with a different solvent wash.

Samples were heated to 70° or 80°C for two hours. The charcoal wires were eluted with 4 ml of purified *n*-pentane and ultrasonic extraction for one minute. The *n*-pentane extracts were concentrated under nitrogen to a final volume of approximately 200 μl prior to gas chromatographic analysis. No interference from water vapor was found. As little as 0.1 μl of gasoline and kerosene, and 0.5 μl of diesel fuel were detected in 1 l cans.

Juhala (1982) also reported success when using carbon coated wires with desorption with carbon disulfide. He also reported using carbon coated Plexiglass® beads for passive adsorption of vapors from fire debris samples. Preparation of the beads is simple, requiring only shaking of the bead with a small amount of charcoal in a dental amalgamator (beads and amalgamator are available from Cresent Dental Company). The beads are placed in the sample container for one hour. The beads are placed in a test tube, and 8 drops of carbon disulfide is used to elute the beads. Similar results to the adsorption wire were obtained with less preparation time required.

Dietz (1991) reported success when using passive sampling techniques with two different charcoal devices. The first involved placing activated charcoal in packets of the porous paper used for tea bags. Some problems were encountered in efficiently desorbing these samples with carbon disulfide. The second and more effective method uses commercially available, carbon-coated Teflon™ strips used in environmental monitoring badges. As little as 0.2 μl of an equal mixture of gasoline, kerosene, and diesel fuel was recovered by using the charcoal strips. Samples were heated at 60°C for one hour or at ambient temperature overnight. No interference from water in the sample was found.

ASTM E: 1413–91, Standard *Practice for separation and concentration of flammable or combustible liquid residues from fire debris samples by passive headspace concentration* offers the analyst a practical guide for using this sample preparation technique.

5.2.3.5 Solvent extraction

Solvent extraction has been used to recover petroleum products from debris longer than gas chromatography has been used as a method for separation and identification of recovered products. Ettling (1963) reported using dichloromethane to extract petroleum products from fire debris. The extracts were analyzed by infra-red spectroscopy. Solvent extraction continued to find use as an isolation technique for gas chromatographic analysis (Ettling & Adams 1968, Midkiff 1978). Midkiff (1978) reported that a variety of solvents were being used including *n*-pentane, hexane, dodecane, hexadecane, methylene chloride, chloroform, carbon tetrachloride, benzene, ethyl ether, and carbon disulfide. The technique is based on the solubility of hydrocarbons found in flammable and combustible liquids in the solvent. Debris samples are typically soaked in a suitable solvent for some time, and the solvent decanted, filtered, and evaporated to a small volume, usually several ml, to concentrate any product isolated prior to instrumental analysis.

An advantage of solvent extraction is the recovery of nonvolatile materials not recovered by vapor sampling techniques. Compounds less volatile than C_{18} are difficult to recover from debris samples by heated headspace or adsorption techniques, both of which depend on the compounds of interest being in the headspace of the debris. Therefore it is not always possible to distinguish between kerosene, which typically ends around C_{17}, and diesel fuel, which extends to C_{23}. Solvent extraction is necessary to distinguish between kerosene and diesel fuel products. Solvent extraction is also the isolation method required for nonvolatile products such as lubricating oils and waxes.

A disadvantage of solvent extraction is that substrate and pyrolysis products are generally soluble and more likely to also be carried over with petroleum products of interest than in headspace or sorption techniques. The presence of these substrate and pyrolysis products complicates the interpretation of the chromatogram. High molecular weight products may also foul injection ports and columns.

ASTM E-1386-90 provides practical guidelines for using solvent extraction to isolate flammable and combustible liquids. The practice is useful for all classes of petroleum products, being useful for distinguishing between various grades of fuel oil, as mentioned earlier. The practice also points out that solvent extraction is useful for sampling nonporous surfaces and small samples.

5.2.3.6 Analysis scheme for analyzing fire debris

The discussion on sample preparation techniques is intended to focus the analyst on the advantages and disadvantages of each technique. No one technique is best for every sample. The sample substrate, and type and quantity of petroleum product that may be present in the debris, must all be considered when deciding how to proceed with the analysis. Based on the advantages and disadvantages of the methods, various analysis schemes involving some or all of the methods can be constructed. Willson (1977) proposed a systematic analysis scheme starting with recovery of the sample to identification of a petroleum product. Following a discussion comparing various recovery methods commonly used today, a systematic analysis scheme will be proposed.

Nowicki & Strock (1983) conducted a comparison study of five different sample preparation techniques. The five techniques were sorbent trap/thermal desorption, charcoal elution, headspace sampling, solvent extraction, and steam distillation. One μl of 80% evaporated gasoline was added to polyurethane foam carpet padding, and the per cent recovered was calculated. No gasoline was recovered by steam distillation. Solvent extraction was the next least efficient. Charcoal trapping and elution with carbon disulfide was 25 times more efficient than headspace and 100 times more efficient than solvent extraction. Sorbent trap with thermal desorption was almost twice as efficient as charcoal trapping, but widely varying results were obtained.

Kubler & Stackhouse (1982) compared dynamic adsorption on charcoal, steam distillation, and heated headspace. On the basis of the recovery of C_{16} from a mixture of the normal alkanes, C_{13}, C_{15}, and C_{16}, they found that heated headspace was less sensitive than steam distillation and dynamic adsorption by a factor of 100. One other important finding of the study was the effect of recovering petroleum products from simulated charred debris. Recoveries of the normal alkanes from charcoal was 10 to 30 times less than from clean Kimwipes™. This is an important point for analysts to remember when evaluating recovery techniques. The sample matrix, especially burned debris, may have a significant effect on the amount of petroleum product that can be recovered. In the case of charred debris, the charred surfaces may tend to retain the petroleum products, making their recovery for analysis more difficult.

Headspace sampling is generally considered to be the least efficient method of recovering petroleum products, and is suitable for screening samples and analyzing samples with an obvious odor, and therefore considerable quantity, of petroleum product. Adsorption/elution on charcoal, either by dynamic or passive adsorption, is a method of choice based on its efficiency and simplicity. The liquid sample obtained from the charcoal can be preserved for later analysis, possibly by an opposing expert. Solvent extraction is needed to distinguish kerosene from diesel fuel products. It may also be the method of choice for nonporous surfaces such as glass from a Molotov cocktail. However, latent fingerprint impressions are destroyed by solvents. Evidence requiring processing for latent fingerprints should not be heated or rinsed with solvent. Both types of evidence, fingerprint and petroleum product identification, might be obtained from glass by rinsing the surface believed to have been exposed to the flammable liquid, such as the inside of a glass bottle, and therefore leaving the side which may contain latent prints untouched by solvents.

Figure 5.10 represents one possible analysis scheme using the more commonly employed sample preparation techniques. The first step involves opening the can to inspect the sample and note any obvious odors. Because of the toxic properties of some fire debris samples, care should be exercised in not exposing oneself directly to the vapors. The small amount of time a container needs to be open for inspection precludes the loss of significant amounts of petroleum product vapor in the headspace. Examining the debris gives the analyst some indication of what kinds of substrate and pyrolysis products may also be recovered. It also avoids problems such as the evidence being double packaged, or containing large quantities of a liquid petroleum product or water, which, on heating, could lead to disastrous results. Some laboratories routinely screen all samples using headspace. If a petroleum product is not identified, another more sensitive technique, such as sorption/elution, is used. Some laboratories skip a headspace screening and go directly to sorption/elution or solvent extraction. One problem with this approach is that products such as alcohols and some light petroleum products may not be detected.

In most cases, headspace methods do not distinguish between Class 4 and Class 5 petroleum products. In fire debris samples, where interactions between petroleum products and charred debris affect the recovery of the hydrocarbon products, compounds less volatile than C_{17} are usually not recovered. Using only headspace recovery, including dynamic and passive adsorption methods, misidentification of a Class 5 product for a Class 4 is possible. Such misidentification was seen in the 1988 and 1989 *Flammables Analysis Proficiency Tests* (Collaborative Testing Services, Inc., 1988, 1989) results where the committee concluded that the reason laboratories misidentified Class 4 products for Class 5 was the sample preparation methods used. Laboratories reporting the Class 4 product used heated headspace and/or adsorption/elution methods. In most cases it is necessary to use solvent extraction to recover compounds less volatile than C_{17}, and therefore distinguish between products such as kerosene and diesel fuel. While the argument can be made that the identification of a petroleum product in the debris is the primary information most investigators require, and distinguishing between Class 4 and 5 is insignificant, analysts must be aware of this limitation of all headspace recovery techniques. Distinguishing between Class 4 and Class 5 may be of significance to an investigation. When a heavy petroleum distillate is detected by any of the headspace methods, the analyst should report that either a Class 4 or Class 5 petroleum product was detected. If solvent extraction is practical for the sample, it may be possible to distinguish between the two classes, since solvent extraction is not limited by the volatility of hydrocarbon compounds. Using the analysis scheme in Fig. 5.11, the analyst can employ a variety of simple recovery techniques to recover and identify most classes of petroleum products.

Another consideration in selecting a sample preparation method is the preservation of the original sample. Proponents of static headspace analysis argue that the original sample is not substantially changed, and that headspace analysis is essentially a nondestructive technique that preserves the sample for subsequent analysis, possibly by an opposing expert. However, the sensitivity limitations of headspace analysis outweigh the benefits of sample preservation. Some sorption methods also preserve the sample for future examination. Thermal desorption methods suffer from the 'one shot' limitation. If all the petroleum product in the sample is sorbed during sampling, no product is left for re-analysis after thermal desorption into the gas chromatograph. Samples which are desorbed with solvent can be preserved for future analysis. It is rarely necessary to consume all the eluent in gas chromatographic analysis. Eluent remaining after analysis can be preserved in a sealed glass container or re-adsorbed onto charcoal where it theoretically could remain stable indefinitely. Likewise, solvent extraction samples can also be preserved for future analysis. Because of the inhomogeneity of fire debris samples, splitting the sample and analyzing only a portion of it is not a practical alternative.

5.2.4 Use of selective detectors

The complexity of most petroleum products recovered from fire debris permits class identification based on pattern recognition. Sometimes more specific information is needed than can be provided from GC-FID patterns alone. Pattern recognition cannot be applied to products which contain single or few peaks. Non-hydrocarbon compounds of interest, such as lead alkyl or oxygenated additives, are not identifiable by a nonspecific detector such as the FID. Volatile materials from the matrix may obscure the hydrocarbon pattern of interest. In these cases, use of a more specific detector can provide information of interest.

5.2.4.1 Other ionization detectors

Photoionization detectors (PID) can be used to discriminate different classes of hydrocarbons. The detector works on the principle that absorption of a photon by a molecule leads to ionization when the photon has an energy greater than the ionization potential of the molecule (Langhorst 1981). By using photons of different energy, the detector can discriminate classes of compounds based on ionization potential differences. Ultraviolet lamps of varying energy are used in commercially available detectors. Driscoll (1982) reported that a 11.7 eV lamp gave responses similar to a FID detector, that is, a universal response to hydrocarbons. Changing to 9.5 eV and 10.2 eV lamps results in discrimination of aromatic, olefin, and aliphatic hydrocarbons. By analyzing a sample, using all three lamps, and measuring the relative peak heights in each chromatogram, some structural information on eluting compounds can be obtained. Discrimination between aliphatic and aromatic hydrocarbons could be useful in the identification of commercial petroleum products.

Another type of ionization detector that has found applications in analysis of petroleum products is the thermionic ionization detector (TID). These detectors originally found use in nitrogen and phosphorus specific detection. The main component of these detectors is an electrically heated thermionic emission source position so that sample compounds impinge on its surface. Ions produced are measured at an adjacent collection plate. Variations in the composition of the emission source, temperature of the source, and gas environment at the source surface provide varying specificity for different compounds (Patterson 1986). Depending on the particular detector configuration used, organolead compounds, halogenated lead scavengers (ethylene dichloride and ethylene dibromide), and a nitro-compound diesel fuel additive can be detected. More recently, a configuration of the TID that specifically detects oxygenates, including simpler alcohols, ketones, MTBE (methyl-*tert*-butyl ether), and higher boiling oxygenates has been developed. These higher boiling oxygenates provide a characteristic chromatographic pattern which appears to allow some discrimination between brands and grades of gasoline (Patterson 1990).

5.2.4.2 Gas chromatography-mass spectrometry

Forensic scientists have recognized the usefulness of gas chromatography-mass spectrometry to provide definitive information on compound identification not provided by GC-FID for many years (Mach 1977, Juhala 1979). However, it was not routinely used in forensic laboratories because the cost of the instrumentation and difficulty of operation precluded routine use in arson investigation. Recent advances in technology have made GC-MS more affordable and simpler to use, and the technique is finding increased use in the analysis of fire debris samples. In addition to identifying samples containing single or few peaks, GC-MS can also be used to identify complicated patterns suspected to be composed only of volatile compounds from the sample matrix. One case report (Howard & McKague 1984) demonstrated the usefulness of GC-MS in identifying peaks in the GC-FID chromatographic pattern of burned carpet material as being the combustion products styrene, benzene and toluene.

Sophisticated, low-cost computerized data systems on mass spectrometers made the analysis of the complex mixtures such as petroleum products feasible. Even with high-speed computation and high resolution separation of capillary columns, it is not feasible to separate and identify each peak on the basis of library searches. Therefore, for use on complex mixtures, some method of extracting information characteristic of petroleum products is necessary. Smith (1983) reported a GC-MS

Table 5.3. Compounds present in petroleum products and characteristic ions†

Compound type	characteristic ion (*m*/*z*)
Saturated aliphatic hydrocarbon	57, 71, 85, 99 (fragments) 100, 114, 128 (molecular)
Alicyclic and unsaturated aliphatic hydrocarbons	55, 69, 83, 97
Alkylbenzenes	91, 105, 119, 133 (fragment) 92, 106, 120 (molecular)
Alkyl-substituted condensed polynuclear aromatics	142, 178 (molecular)
Monoterpenes	93 (fragment) 136 (molecular)

†R. Martin Smith (1983) 'Mass chromatographic analysis of arson accelerants', *Journal of Forensic Sciences* **28** (2), 318–329.

data analysis method that identified compounds characteristic of petroleum products. This method is mass chromatography. Smith found that several characteristic ions for the major classes of compounds present in petroleum products could be identified. These classes and characteristic ions are listed in Table 5.3. Mass chromatograms corresponding to each family of compounds were calculated by summing the intensities of up to four ions characteristic of the group. By comparing unknown to standard mass chromatograms for identification of types of compounds and crude quantitation, the presence of a class of petroleum product can be determined. The additional chemical information provided by GC-MS can aid the analyst in interpretation of inconclusive GC-FID patterns. For a comprehensive overview of the use of mass chromatography in arson analysis, the reader is referred to a recent review by Smith (1987).

Kelly & Martz (1984) reported a similar treatment for GC-MS analysis for petroleum products. In addition to identifying classes of petroleum products, the utility of the technique for identifying additives such as methyl-*tert*-butyl ether (MTBE) and lead alkyl compounds was demonstrated. Bertsch *et al.* (1988) reported a data analysis scheme using computerized data extraction and analysis that summarizes qualitative and quantitative information on samples in a one-page summary report. This eliminates manual data entry and manipulation by the analyst, simplifying the interpretation process.

Nowicki (1990) has proposed using GC-MS identification of classes of compounds as the basis of an accelerant classification scheme. This scheme is based on chemical composition identified by mass spectrometry in addition to the boiling point range information as shown in the classification scheme in Table 5.1. Fragment ions are identified that provide specific analyses for alkanes, cyclo-paraffins, aromatics, naphthalenes, and dihydroindenes. The expanded accelerant classification scheme takes into account the chemical nature of petroleum products in addition to boiling point ranges.

Another approach to GC-MS data interpretation is based on identification and quantitation of target compounds (Fultz & Wineman 1990, Keto & Wineman 1991).

Target compounds characteristic of petroleum products in gasoline, medium range petroleum products, kerosene, and diesel fuel were identified. GC-MS is used to qualitatively and semi-quantitatively identify the target compounds present in samples isolated from fire debris. Quantitative data are converted into a stick plot, termed a target compound chromatogram (TCC). The TCC shows the identity, relative abundance, and retention time of each targeted compound found in the sample. The TCC of the unknown can be compared to known standard without interference from pyrolysis products, enhancing the identification of petroleum products. A study of pyrolysis products from common fire debris matrices, such as wood, carpeting, and floor tile, showed that some targeted compounds were detected in the pyrolyzed isolates of these materials. However, the quantities were different from those found in fire debris samples, and the TCC of pyrolysis products are readily distinguishable from those of petroleum products.

5.2.4.3 Gas chromatography-Fourier transform infra-red spectroscopy

While the mass spectrometer seems to be the most promising detector for providing more specific information on petroleum products in fire debris samples, one application of Fourier Transform Infrared Spectroscopy (FTIR) to the identification of petroleum products in fire debris samples has been published (Hipes *et al.* 1991). Samples of gasoline, typical light and mid-range petroleum products, and kerosene were analyzed by GC-FTIR in an effort to determine if characteristic IR absorption bands could be identified and linked to each type of product. Only gasoline showed a strong enough absorption in the 3080–3010 cm^{-1} spectral window (aromatic ring C-H stretches) to be useful for detection. Light and mid-range products, along with kerosene, being composed of mainly aliphatic hydrocarbons, had no unique absorption bands in the infra-red. Experiments were also done to determine if FTIR could distinguish petroleum products from background materials such as terpenes and styrenes. GC/MS using selected ion monitoring was found to be more discriminating than any infra-red examination. In a comparison of the relative sensitivities of the IR, MS, and FID detectors used with GC, the FTIR was the least sensitive, followed by MSD, with FID being the most sensitive. The technique did not perform as well as GC/MS in detecting petroleum products in the presence of common pyrolysis products or other types of petroleum products.

5.3 GAS CHROMATOGRAPHY IN EXPLOSIVES ANALYSIS

5.3.1 Introduction

Prior to the introduction of fused silica capillary columns, the thermal lability and polarity of most explosives made them unsuitable for trace analysis by gas chromatography because of degradation in heated injection ports and irreversible adsorption on some chromatographic columns. In addition, unlike petroleum products, explosives residues are composed of only one or at most a few chemical compounds. Therefore, chromatographic characterization is limited to retention time rather than a pattern of multiple peaks which can be used in pattern recognition. The

Table 5.4. Commercial and military explosive compounds

Abbreviation	Name(s)	Molecular formula
DNT	Dinitrotoluene isomers	$C_6H_4CH_3(NO_2)_2$
EGDN	Ethylene glycol dinitrate	$(CH_20)_2(NO_2)_2$
HMX	1,3,5,7-tetranitro-1,3,5,7-tetrazacyclooctane Octagen	$C_4H_8N_8O_8$
ISD	Isosorbitol dinitrate	$C_6H_8N_2O_8$
NC	Nitrocellulose, cellulose nitrate	$[C_6H_7N_3O_{11}]_n$
NG	Nitroglycerin, glycerol trinitrate 1,2,3-Propanetriol trinitrate	$C_3H_5N_3O_9$
PETN	Pentaerythritol tetranitrate	$C_5H_8N_4O_{12}$
RDX	1,3,5-trinitro-1,3,5-triazacyclohexane Hexagen, Cyclonite	$C_3H_6N_6O_6$
Tetryl	2,4,6-N-tetranitro-N-methylaniline Nitramine	$(NO_2)_3C_6H_2N(CH_3)NO_2$
TNT	Trinitrotoluene	$C_6H_3CH_3(NO_2)_3$
Non-explosive ingredients		
DPA	Diphenylamine, N-phenylbenzeneamine	$C_6H_5NHC_6H_5$
EC	Ethyl centralite	$C_{17}H_{20}N_2O$

analysis of explosives compounds requires a selective detector so that identification can be made on chemical properties. An early review on the analysis of explosives (Yinon 1977) reported that gas chromatography could be used for the analysis of bulk explosives, but the technique was more amenable to nitroaromatics for the stability reasons mentioned above. Table 5.4 lists explosive compounds which have been analyzed by gas chromatography. Packed column gas chromatography with FID detection was used to separate mono- and di-nitrotoluene isomers, detect TNT in cyclotol (a mixture of RDX and TNT), nitroglycerin in double-base smokeless powders, and some plasticizers and stabilizers in propellants. With the use of more sensitive detectors, some laboratories now include gas chromatographic separation as a part of their analysis schemes for the identification of explosive compounds (Rudolph & Bender 1983, Garner *et al.* 1986). The use of on-column injection can also be used to eliminate adsorption of polar compounds on hot glass liners or heated needles (Penton 1983). Sub-picogram quantities of nitroglycerin were detected by using on-column injection to a fused silica capillary column, with an electron capture detector.

Irrespective of the detector used, prior to the introduction of commercially available inert fused silica capillary columns, the analysis of explosives by gas chromatography was limited and sometimes frustrated by the lability and adsorptivity properties of these compounds. The problems of gas chromatographic separations can be circumvented, and several gas chromatographic methods have been applied to the analysis of explosive compounds. In addition to the need for a selective detector, trace analysis of explosives in environmental and post-blast debris samples requires a detector considerably more sensitive than the FID.

Beveridge (1986) reviewed the analysis of explosives residues, using a variety of analytical methods. While gas chromatographic analysis with specific detection can be used for screening purposes, the other methods are often employed to confirm the presence of trace amounts of explosives in debris samples. This section will concentrate on the post-blast analysis of explosives, where the small quantities of explosives in residue require selective and sensitive methods capable of separating the compounds of interest from other materials in the sample.

Capillary supercritical fluid chromatography has been used to separate a variety of explosive compounds (Munder *et al.* 1990). Using both an UV and a FID detector, these workers found that the elution order using carbon dioxide as a mobile phase on a nonpolar column was similar to that achieved by gas chromatography. Supercritical fluid chromatography, which does not require volatilization of the sample for analysis, and shows improved separation over conventional HPLC, may be a useful technique for the separation and detection of thermally labile explosive compounds.

5.3.2 Electron capture detection

The electron capture detector was the first to provide the selectivity and sensitivity necessary for detection of trace amounts of explosives. A portable GC-ECD instrument was developed in 1978 (Elias) to sample the cabin of aircraft for explosive vapors. The prototype was developed to detect EGDN, a component of most dynamites. Air samples were preconcentrated on solid sorbent prior to injection into the portable instrument. Sampling for thirty minutes, concentrations down to 0.05 ppt of EGDN were detected. An obvious limitation of this type of detection is the required volatility of the compounds being detected. Commercial explosives such as slurries and gels, and military explosives such as C-4, TNT, and Semtex, would not be amenable to this type of detection.

Yandler & Hanley (1978) reported collecting vapors of the nitrate esters nitroglycerin and ethylene glycol dinitrate on the porous polymer Porapak P.S. (100/120 mesh) using a portable vacuum pump at a rate of 5–6 ml/min. Headspace samples were injected into a GC with a six foot × 2 mm ID glass column packed with 3% OV-210 coated Chromosorb. The carrier gas was argon/methane 95:5, and a tritium (^{3}H) source ECD was used for detection. In post-blast field tests of dynamite detonations, EGDN was detected, but not NG. The authors speculate that the lower vapor pressure of NG combined with low sampling rate may account for the lack of detection of NG. Samples packaged and transported to the laboratory for analysis yielded better recoveries of explosive vapors.

To circumvent the adsorption problems encountered with packed columns, Douse (1981) used silica capillary columns coated with OV-101 to show that these columns could separate picogram quantities of explosives. A ^{3}H electron capture detector was used. On the basis of work by previous workers reporting the limit of detection of explosives on packed columns to the low nanogram levels, Douse studied leached persilylated soda glass, barium carbonate coated silica glass, and untreated silica glass coated with a nonpolar stationary phase. Using standard solutions of PETN (pentaerythritol tetranitrate), Tetryl, HMX, NG (nitroglycerin), TNT (trintrotoluene), and RDX, Douse separated and detected 100 picogram quantities of these compounds on the 25 m by 0.25 mm ID 0V-101 column. Douse (1982) used this same method to analyze handswab extracts for the presence of explosive compounds. A pre-analysis clean-up using porous polymer beads was developed to remove interfering compounds prior to GC analysis. Nanogram quantities of explosives were detected. Jane *et al.* (1983) used this GC-ECD method to identify nitroglycerin as part of an overall analysis scheme for the identification of organic constituents of gunshot residue. Thin-layer chromatography was used for detection of nitrocellulose and nitroglycerine, and HPLC was used for the detection of diphenylamine.

Douse & Smith (1986) reported an improved method for the isolation and purification of organic residues from firearm discharges taken by handswabs. These extracts were analyzed by capillary GC using a 12 m × 0.25 mm ID BP-1 fused silica capillary column, 0.25 μm film thickness. A ^{63}Ni electron capture detector was used. The authors reported that significant amounts of co-extractives were also detected by the ECD in very dirty handswab extracts. NG, TNT, and RDX in the low nanogram range were detectable in spiked handswab extracts.

Yip (1982) employed short, mixed liquid phases and combined packed-capillary columns to separate EGDN (ethylene glycol dinitrate), EGMN (ethylene glycol mononitrate), and NG at levels from 10^{-12}–10^{14} g/ml. A short packed column (<0.6 m) with a mixed stationary phase of 1:1 OV-101 and OV-17 was found to be suitable for screening vapor samples of debris for these explosives. A ^{63}Ni electron capture detector was used. The combination of short columns loaded with 5% stationary phase on highly inert packings combined with low operating temperatures and high carrier gas flows prevented thermal decomposition and adsorption of the compounds prior to detection.

Krull *et al.* (1983) reported a method using packed column separation with dual detection using an ECD and photoionization detector (PID). The ECD and PID have significantly different response factors from nitrated compounds. By calculating the ratio of the response to a compound in each detector, a relative response factor characteristic of a compound is obtained. This relative response factor can be used to differentiate compounds of interest from contaminants recovered from complex sample matrices. While most of the study discusses the use of the technique for the analysis of nitrated aromatic, aliphatic, and polynuclear aromatic compounds, the authors demonstrated the use of ECD/PID detection for identification of TNT, RDX, Tetryl, and NG. A 6 ft × 2.0 mm glass column packed with Permabond Methyl Silicone with nitrogen as the carrier gas was used to separate the compounds. TNT and RDX were not detected by the PID at microgram levels or below. Both detectors responded to Tetryl and NG, and characteristic relative response factors can be calculated for these compounds.

5.3.3 Nitric oxide selective detector

The electron capture detector provided analysts with the selectivity and sensitivity needed to determine trace amounts of explosive compounds. However, in complex sample matrices, even this detector did not provide sufficient specificity for identification of trace amounts of explosives. The ECD can suffer from overloading and contamination, with subsequent loss of sensitivity.

The TEA Analyzer™, a detector originally developed for the detection of nitrosamines, which are animal carcinogens, has found use as a detector for nitrogen-containing explosive compounds (Lafleur *et al.* 1978). This nitric oxide selective pyrolysis/chemiluminescence detector works by passing effluent from the gas chromatograph into a pyrolyzer where NO_2 is released from organic nitro compounds and converted to NO by a catalytic surface. The NO gas is reacted with ozone (O_3) at reduced pressure. This reaction produces a characteristic infra-red chemiluminescence which is detected by a photomultiplier tube. The detector is highly selective for nitro-containing compounds, and is sensitive at the picogram level (Goff *et al.* 1983).

Douse (1985) compared the ECD with the TEA™ for trace analysis of explosive compounds in handswab extracts. He found the TEA™ to be more selective and sensitive than the ECD, enabling low nanogram levels of explosives in handswabs to be detected. In a comparison of the thermionic specific detector (TSD), ECD, Hall electrolytic conductivity detector with the TEA™ for the gas chromatographic analysis of nitroaromatics in biosludges, the TEA™ exhibited the greatest selectivity (Phillips *et al.* 1983). All these detectors are capable of measuring nitroaromatics in the picogram range and are selective for nitroaromatics versus hydrocarbons. Interfering organic compounds extracted from the sludge were detected by the TSD, ECD, and HECD that were not detected by the TEA™.

The TEA™ can be used as a detector for either gas chromatography or high performance liquid chromatography. Lafleur *et al.* (1978) reported detection of nanogram quantities of EGDN, RDX, HMX, PETN, and Tetryl when using HPLC-TEA™. Gas chromatographic separation did not give as satisfactory results. At that time, the thermal lability and adsorptivity of some explosive compounds had not been circumvented by using fused silica capillary column technology. Lafleur & Mills (1981) reported improved selectivity for nitroaromatic compounds using GC-TEA™ over HPLC-TEA™. Nitroaromatic compounds are amenable to gas chromatographic analysis, even on packed columns. A glass 1.8 m × 2.0 mm ID column packed with 3% OV-225 on Chromosorb W-HP 100/120 mesh, with argon as a carrier gas at 30 ml/minutes, was used for the separation. The detector gave a linear response for a variety of nitrotoluenes over four orders of magnitude and detected standard solutions at the picomole level.

Goff *et al.* (1983) reported low picogram detection of NG, PETN, Isosorbide dinitrate (ISDN), EGDN, 2,4-DNT, TNT, RDX, and Tetryl, using on-column injection to a fused silica capillary column (DB-5). The 30 m × 0.32 mm column had a film thickness of 0.25 μm; helium carrier gas at an 18 psi head pressure was used. The injection port was held at ambient temperatures, and an oven temperature program of 60°C for one minute, ramping at 15°C/minute to 240°C, holding for three minutes, was used to separate explosive compounds of interest. The low picogram sensitivity was maintained even in complex samples because of the selectivity of the detector. This eliminates the need for elaborate sample clean-up prior to analysis. The authors recommended using HPLC-TEA™ as a confirmatory approach.

The use of this technique on 'real-world' samples was demonstrated by Fine *et al.* (1984). Post-blast debris from field tests on C-4 (containing RDX), TNT, and detonating cord (containing PETN) were analyzed with the above protocol. Sub-nanogram quantities of the explosives were detected by GC and HPLC. Another experiment involved taking handswabs from volunteers handling gelatin dynamite or C-4. EGDN and NG were easily detected from the swabs taken from volunteers handling dynamite. RDX was detected in the swabs from handlers of C-4. Because of the selectivity of the detector, no pre-analysis clean-up was necessary. The minimum detectable amount of most explosives was reported to be 4–5 picograms injected on-column.

Douse (1983) also reported the detection of explosive compounds from handswabs in the low picogram levels, using fused-silica capillary columns and the TEA™ detector. The minimum detectable quantities were not as low as reported elsewhere (Fine 1983, Goff *et al.* 1983), and it should be noted that heated splitless injection at 200°C was used rather than ambient on-column injection. Douse also reported that sensitivity was improved by inserting the end of the capillary deep into the heated zone of the pyrolysis tube as opposed to the transfer line. Poor results for tetryl and PETN were obtained, and this was attributed to the use of too thick a film on the column (stationary phase OV-101 film thickness not noted). Douse (1987) reported improved detection by altering the detector, substituting the ceramic pyrolysis tube with fused silica and eliminating the cold trap. Collins (1989) reported further improvement in the chromatography of explosives with a TEA™ by modifying the circuitry to improve response time to take full advantage of the resolving power of capillary gas chromatography.

Lloyd (1991) reported a variation on the combined use of HPLC and GC with TEA™ for the analysis of explosives recovered from handswabs. Citing the need to confirm the results of one separation method, Lloyd developed a method to trap HPLC peaks by using a microcolumn. These trapped peaks, tentatively identified by using a pendant mercury drop electrode detector, are then injected into a gas chromatograph-TEA™. The microcolumn eluents are injected directly onto a GC retention gap of unmodified silica, eliminating the need of evaporative concentration prior to GC analysis.

5.3.4 Mass spectrometry detector

Mass spectrometry has found use as a research tool and method of forensic analysis for explosive compounds. A great deal of work has been done on studying explosives by mass spectrometry alone without any chromatographic separation prior to analysis (Yinon 1982). Analysis of explosives in complex sample matrices necessitates a separation procedure prior to mass spectral analysis. Both gas and liquid chromatography have been used to separate compounds of interest prior to mass spectroscopic analysis. The advent of inert fused silica capillary columns brought renewed interest to the use of GC/MS for the analysis of explosive compounds.

Electron impact mass spectrometry (EIMS), which uses an electron energy of about 70 eV, produces extensive fragmentation of explosive compounds, rarely yielding a molecular ion fragment. These fragments give functional group information, and can produce an overall pattern characteristic of the compound. However, these patterns are not useful in mixtures, since the resulting patterns are too complex for interpretation. With chromatographic separation prior to analysis, these patterns can be useful for identification and detection of explosive compounds. Chemical ionization (CI), a soft ionization technique, produces a fragmentation pattern that often includes a molecular ion. Combined with EI fragmentation patterns, the two techniques can provide a reliable identification tool for explosives. For a thorough review of the forensic applications of mass spectroscopic analysis of explosives, readers are referred to Yinon (1987).

Pate & Mach (1978) reported the separation and identification of EGDN, NG, 2,4-DNT, and TNT at the nanogram level on glass packed columns, using CIMS. Nitrate esters were separated on 3% OV-1 on 100/120 mesh Chromosorb W, while nitroaromatics were run on 3% OV-17 on 80/100 Supelcoport. The authors note that the analysis of the nitrate esters are difficult because of their thermal lability and tendency to decompose on hot metal surfaces. Electron impact mass spectra of nitrate esters gives NO^{2+} as the base peak, with negligible amounts of higher weight ions that could be used to differentiate the various compounds. Methane chemical ionization gave characteristic higher weight ions along with pseudomolecular ions that could be used to differentiate the nitrate esters. EGDN, NG, and PETN gave ions at 153, 228 and 317 m/z, respectively. It should be noted that PETN, nitramines (tetryl and RDX), and ethyl centralite were not amenable to gas chromatographic introduction into the mass spectrometer, and were introduced by solid probe.

CIMS was used to characterize organic constituents from gunshot residue (Mach *et al.* 1978a,b). Using powder from commercial ammunition, up to five organic compounds were identified by GC/MS: NG, 2,4-DNT, DPA, DBP (dibutylphthalate), and EC (ethyl centralite). The analysis was done on short packed columns, using CI ionization with methane gas. One important finding of this study was that 'brands' of ammunition could not be identified from profiles of organic constituents identified in the 33 powders studied. Cartridge manufacturers use different powders, even in the same caliber ammunition, to achieve performance requirements. One company may purchase a lot of powder from another, and rework it, adding explosive components and stabilizers, to meet its own requirements. The situation is similar to attempts to identify brands of gasoline. Because of the marketing and manufacturing processes of the industry, the identification of a particular brand from the unburned powder left behind on a target cannot be done.

In a study recovering gunshot residues from actual test fires (Mach *et al.* 1978b), large particles of smokeless powder can persist onthe hands no more than one hour after firing. The study also found that there is considerable cross-contamination of organic constituents from different types of ammunition fired in the same gun, even when the gun is cleaned between test fires. No volatile organic constituents were recovered from handswabs from which portions of original smokeless powder flakes were excluded. While the sensitivity of the GC/CIMS was determined to be in the nanogram range from limited ion monitoring and picogram range for selected ion monitoring, interference from skin oils appeared to preclude analysis by GC/CIMS.

Using a novel data processing and searching approach, Martz *et al.* (1983)

developed a reloader smokeless powder library based on electron impact mass spectral data of the volatile components of smokeless powders. Chloroform extracts of smokeless powder particles were injected on-column at ambient temperatures on a 20 m × 0.2 mm ID SE-54 bonded phase fused silica capillary column. Scanning from 45 to 400 AMU, total ion chromatogram data were collected for each powder. The data processing entailed summing the mass spectra of all the components of interest in the powder and substracting any contribution from background. While the EI spectrum of any component alone is not characteristic, the summed spectra are characteristic of a particular smokeless powder. The summed spectra contain information from both components and relative quantities of these components in the powders. Over 80 reload powders have been characterized by using this approach. The technique has been used to identify brands of smokeless powders recovered from improvised explosive devices, and to compare smokeless powder deposits on gunshot victims with residues from fired cartridge cases. The authors note that on-column injection is the key to the successful collection of data, since it allows introduction of components onto the GC column without degradation.

Nitrate esters in blood plasma have been successfully detected and quantitated by using electron capture negative ion mass spectrometry prior to gas chromatographic separation (Bignall *et al.* 1981). This ionization method produces an ion at m/z 62 that is suitable for single ion monitoring of NG, EGDN, ISD, and PETN, giving lower detectable limits for NG of 80 pg/ml extracted from blood. Short narrow bore packed columns were used for the separation, and the authors note that degradation of the compounds during gas chromatographic analysis was the major limitation of the technique.

Cumming & Park (1983) also took advantage of the strong electron capturing property of explosive compounds to produce negative ion mass spectra and thereby improve detection limits and ability to discriminate the various compounds. Detection limits for EGDN, NG, PETN, RDX, and TNT were compared, using positive ion operation, were chromatographed on three different columns, packed, borosilicate capillary, and fused silica capillary. Heated injection was used for all samples. Fused silica capillary columns provided the greatest sensitivity for all explosives analyzed. The combination of EI and NI spectra along with GC retention time gives sufficient information for identification of explosive compounds at minimum levels between 100–150 ng. Single ion monitoring can reduce minimum detectable amounts to picogram levels.

GC/MS using EI or CI is used to confirm post-blast identification of explosives tentatively identified by thin-layer chromatography (Tamiri & Zitrin 1986, Zitrin 1986). A 15 m × 0.25 mm ID DB-5 (0.25 μm film thickness) fused silica column with heated injection was used to separate TNT, NG, EGDN, and RDX at 10 ng levels. EI spectra are more helpful when prior indication of the identity of the explosive by another analytical technique, such as thin-layer chromatography, has been obtained. As pointed out by other workers, the authors found that CI spectra are more useful in identification of nitrate esters. Identification of plasticizers and stabilizers can be useful for distinguishing NG from dynamite as opposed to NG from double-base smokeless powder. The identification of ethyl centralite in post-blast debris from a car bombing differentiated the source of detected NG as being from smokeless powder rather than dynamite.

REFERENCES

AA Notes (1982), **6** (3) 57–59.

Aldridge, T.A. (1981), A thin layer chromatographic clean-up for arson distillates. *Arson Analysis Newsletter* **5** (3) 39–42.

Aldridge T.A. & Oates, M. (1984), Fractionation of accelerants and arson residues by solid phase extraction. *Journal of Forensic Sciences* **31** (2) 666–686.

Andrasko, J. (1983), The collection and detection of accelerant vapors using porous polymers and curie point pyrolysis wires coated with active carbon. *Journal of Forensic Sciences* **28** (2) 330–344.

Armstrong, A.T. & Wittkower, R.S. (1978), Identification of accelerants in fire residues by capillary column gas chromatography. *Journal of Forensic Sciences* **23** (4) 662–671.

ASTM E-1412-91, Standard Practice for separation and concentration of flammable or combustible liquid residues from fire debris samples by dynamic headspace concentration.

ASTM E-1413-91, Standard Practice for separation and concentration of flammable or combustible liquid residues from fire debris samples by passive headspace concentration.

ASTM E-1385-90, Standard Practice for separation and concentration of flammable or combustible liquid residues from fire debris samples by steam distillation.

ASTM E-1386-90, Standard Practice for separation and concentration of flammable or combustible liquid residues from fire debris samples by solvent extraction.

ASTM E-1387-90, Standard Test Method for Flammable or Combustible Liquid Residues in Extracts from Samples of Fire Debris by Gas Chromatography.

ASTM E-1388-90, Standard Practice for sampling of headspace vapors from fire debris samples.

ASTM E-1389-90, Standard Practice for clean-up of fire debris sample extracts by acid stripping.

Baldwin, R.E. (1977), Adsorption-elution technique for concentration of hydrocarbon vapors. *Arson Analysis Newsletter* **1** (6) 9–12.

Bertsch, W., Sellers, C.S., Babin, K. & Holzer, G. (1988), Automation in the chemical analysis of suspect arson samples by GC/MS. A systematic approach. *Journal of High Resolution Chromatography & Chromatography Communications* **11** 815–819.

Bertsch, W. & Zhang, Q. (1990), Sample preparation for the chemical analysis of debris in suspect arson cases. *Analytica Chimica Acta* **236** 183–195.

Beveridge, A. (1986), Explosives residue analysis in the mid-1980's – an expanding and challenging role for forensic scientists. *Journal of Energetic Materials* **4** 29–75.

Bignall, J.C., Davies, N.W., Power, M., Roberts, M.S., Cossum, P.A. & Boyd, G.W. (1981), The analysis of nitrate esters by combined gas chromatography and electron capture negative ion mass spectrometry. *Recent Developments in Mass Spectrometry in Biochemistry, Medicine and Environmental Research* **7** 111–122.

Brackett, J.W., Jr. (1955), Separation of flammable material of petroleum origin from evidence submitted in cases involving fires and suspected arson. *Criminal law, Criminology and Police Science* **46** 554–561.

Brettell, T.A., Moore, P., La Machia, M. & Grob, R. L. (1984), The detection of arson accelerants using headspace gas chromatography and a wide-bore capillary column. *Arson Analysis Newsletter* **8** (4) 86–107.

Brettell, T.A., Moore, P.A. & Grob, R.L. (1986), Detection of arson accelerants using dual wide-bore bonded-phases capillary columns and static headspace sampling. *Journal of Chromatography* **358**, 423–428.

Caddy, B., Smith, F.P. & Macy, J. (1991), Methods of fire debris preparation for detection of accelerants. *Forensic Science Review* **3** (1) 58–69.

Chrostowski, J.E. & Holmes, R.N. (1979), Collection and determination of accelerant vapors from arson debris. *Arson Analysis Newsletter* **3** (5) 1–16.

Clark, H.A. & Jurs, P.C. (1975), Qualitative determination of petroleum sample type from gas chromatograms using pattern recognition techniques. *Analytical Chemistry* **47** 374–378.

Clark, H.A. & Jurs, P.C. (1979), Classification of crude oil gas chromatograms by pattern recognition techniques. *Analytical Chemistry* **51** 616–623.

Clodfelter, R.W. & Hueske, E.E. (1977), A comparison of decomposition products from selected burned materials with common arson accelerants. *Journal of Forensic Sciences* **22** (1) 116–118.

Collaborative Testing Services, Inc. (1988), *Flammables analysis*, Report No. 88–9. Crime Laboratory Proficiency Testing Program, Herndon, VA.

Collaborative Testing Services, Inc. (1989), *Flammables analysis*, Report No. 89–8. Crime Laboratory Proficiency Testing Program, Herndon, VA.

Collins, D.A. (1989), Modification to a thermal energy analyzer with associated electronic filtering for improved gas chromatographic analysis of explosives traces. *Journal of Chromatography* **483** 379–383.

Conrad, F.J., Burrows, T.A. & Williams, W.D. (1979), Selection of a gas chromatographic material for use in explosives vapor preconcentration. *Journal of Chromatography* **176** 37–41.

Crandall, D.S. & Pennie, J.T. (1984), A modified technique for the collection of volatiles from fire debris. *Arson Analysis Newsletter* **8** (2) 47–49.

Cummings, A.S. & Park, D.P. (1983), The analysis of trace levels of explosives by gas chromatography/mass spectrometry. *Proceedings of the International Symposium on the Analysis and Detection of Explosives* March 29–31, US Department of Justice, Federal Bureau of Investigation, 259–266.

DeHaan, J.D. (1978), Arson Evidence Packaging. *Arson Analysis Newsletter* **2** (3) 9–13.

DeHaan J.D. (1991), *Kirk's Fire Investigation* 3rd ed. Brady Fire Sciences Series, Prentice-Hall, Inc.

DeHaan, J.D. & Bonarius, K. (1988), Pyrolysis products of structure fires. *Journal of the Forensic Science Society* **28** (5/6) 299–309.

DeHaan, J.D. & Skalsky, F.A. (1981), Evaluation of KAPAK plastic pouches. *Arson*

Analysis Newsletter **5** (1) 6–11.

Dietz, W.R. (1991), Improved charcoal packaging for accelerant recovery by passive diffusion. *Journal of Forensic Sciences* **35** (1) 111–121.

Dietz, W.R. & Mann, D.D. (1988), Evidence Contaminated by polyester bags. *Scientific Sleuthing Newsletter* **12** (3) 5–6.

Douse, J.M.F. (1981), Trace analysis of explosives at picogram levels with silica capillary column gas-liquid chromatography and electron capture detection. *Journal of Chromatography* **208** 83–88.

Douse, J.M.F. (1982), Trace analysis of explosives in handswab extracts using amberlite XAD-7 porous polymer beads, silica capillary column gas chromatography with electron-capture detection and thin-layer chromatography. *Journal of Chromatography* **234** 415–425.

Douse, J.M.F. (1983), Trace analysis of explosives at the low picogram level using silica capillary column gas chromatography with thermal energy analyser detection. *Journal of Chromatography* **256** 359–352.

Douse, J.M.F. (1985), Trace analysis of explosives at the low nanogram level in handswab extracts using columns of amberlite XAD-7 porous polymer beads and silica capillary column gas chromatography with thermal energy analysis and electron-capture detection. *Journal of Chromatography* **328** 155–165.

Douse, J.M.F. (1987), Improved method for the trace analysis of explosives by silica capillary column gas chromatography with thermal energy analysis detection. *Journal of Chromatography* **410** 181–189.

Douse, J.M.F. & Smith, R.N. (1986), Trace analysis of explosives and firearm discharge residues in the metropolitan forensic science laboratory. *Journal of Energetic Materials* **4** 169–186.

Driscoll, J.N. (1982), Identification of hydrocarbons in complex mixtures using a variable energy PID and capillary column gas chromatography. *Journal of Chromatographic Science* **20** 91–94.

Drysdale, D. (ed.) (1985), *Au Introduction to Fire Dynamics*, John Wiley, New York.

Dyroff, G.V. (ed.) (1989), *Manual on Significance of Tests for Petroleum Products* Philadelphia, PA: ASTM 3.

Elias, L. (1978), Recent project at NRC related to explosives detection. *Proceedings: New concepts symposium and workshop on detection and identification of explosives,* October 30–November 1, 1978, 265–267, NTIS: Springfield, VA.

Ettling, B.V. (1963), Determination of hydrocarbons in fire remains. *Journal of Forensic Sciences* **8** 261–267.

Ettling, B.V. (1974), Analysis of paraffin wax in fire remains. *Journal of Forensic Sciences* **20** (3) 476–483.

Ettling, B.V. & Adams, M.F. (1968), The study of accelerant residues in fire remains. *Journal of Forensic Sciences* **13** (1) 76–89.

Fine, D.H., Yu, W.C., Goff, E.U., Bender, E.C., Reutter, D. (1984), Picogram analysis of explosive residues using the TEA analyzer. *Journal of Forensic Science* **29** (3) 732–746.

Forensic Science & Engineering Committee, International Association of Arson Investigators (1988) Guidelines for laboratories performing chemical and instrumental analyses of fire debris samples. *Fire and Arson Investigator* **38** (4) 45–48.

Frenkel, M., Tsaroom, S., Aizenshtat, Z., Kraus, S. & Daphna, D. (1984), Enhanced sensitivity in analysis of arson residues: an adsorption-tube/gas chromatograph method. *Journal of Forensic Sciences* **29** (3) 723–731.

Fultz, M.L. & Wineman, P.L. (1988), Effects of sampling methods on organolead compounds. Presented: *40th Annual Meeting, American Academy of Forensic Sciences, Philadelphia, PA.* February, 1988.

Fultz, M.L. & Wineman, P.L. (1990), A GC/MS procedure for detecting petroleum-derived accelerants in high background fire debris isolates. Presented: *42nd Annual Meeting, American Academy of Forensic Sciences, Cincinnati, OH.* February, 1990.

Fultz, M.L., Ventuarelle, C.A. & Ford, L.C. (1991), Evaluation of the use of commercially available sorbent material for sample collection at fire scenes. Presented: *43rd Annual Meeting, American Academy of Forensic Sciences. Anaheim, CA.* February, 1991.

Garner, D.D., Fultz, M.L., Byall, E.B. (1986), The ATF approach to post-blast explosives detection and identification. *Journal of Energetic Materials* **4** 133–148.

Garten, M.A. (1982), Capillary gas chromatography of isoparaffinic accelerants. *Arson Analysis Newsletter* **6** (4) 65–71.

Gibbs, G.B. & Hoffman, H.L. (1968), Petroleum products. Kirk/Othmer, *Encyclopedia of Chemical Technology* 2nd ed. **15** John Wiley & Sons, Inc.

Goff, E.U., Yu, W.C., Fine, D.H. (1983), Description of a nitro/nitroso specific detector for the trace analysis of explosives. *Proceedings of the International Symposium on the Analysis and Detection of Explosives* March 29–31, US Department of Justice, Federal Bureau of Investigation, 159–168.

Guinther, C.A., Jr., Moss, R.D. & Thaman, R.N. (1983), The analysis and identification of weathered or fire-aged gasoline at various stages of evaporation. *Arson Analysis Newsletter* **7** (1) 1–5.

Henderson, R.W. & Lightsey, G.W (1984), *Fire and Arson Investigator* **35** 8.

Hipes, S.E., Witherspoon, J.W., Bertleson, G.A., Beyer, P.A. & Kurz, M.E. (1991), Evaluation of GC-FTIR for the analysis of accelerants in the presence of background matrix materials. *MAFS Newsletter* **20** (1) 48–76.

Howard, J. & McKague, A.B. (1984), A fire investigation involving combustion of carpet material. *Journal of Forensic Sciences* **29** (3) 919–922.

Hrynchuk, R., Cameron, R. & Rodgers, P.G. (1977), Vacuum distillation for the recovery of fire accelerants from charred debris. *The Canadian Society of Forensic Science Journal* **10** (2) 41–50.

Hurteau, W. (1973), The Arson Evidence Package. *Fire Journal* July, 47–49, 54.

Iglauer, N. (1972), *Pyrolysis-gas chromatography – a method for rapid identification of polymers.* Technical Report, Task 736702, Air Force Material Laboratory, October.

Irwin, W.J. (1985), *Analytical Pyrolysis*, Marcel Dekker, Inc., 134.

Jane, I., Brookes, P.G., Douse, J.M.F. & O'Callaghan, K.A. (1983), Detection of gunshot residues via analysis of their organic constituents. *Proceedings of the International Symposium on the Analysis and Detection of Explosives* March 29–31, US Department of Justice, Federal Bureau of Investigation, 475–483.

Juhala, J.A. (1979), Determination of fire debris vapors using an acid stripping procedure with subsequent gas chromatographic and gas chromatography/mass spectrometry analysis. *Arson Analysis Newsletter* **3** (4) 1–19.

Juhala, J.A. (1982), A method for adsorption of flammable vapors by direct insertion of activated charcoal into the debris samples. *Arson Analysis Newsletter* **6** (2) 32–36.

Juhala, J.A. & Beever, F.K. (1986), A sensitive, rapid & economical method for analyzing fire debris using adsorption elution techniques. *Arson Analysis Newsletter* **9** (1) 1–11.

Kelly, R.L. & Martz, R.M. (1984), Accelerant identification in fire debris by gas chromatography/mass spectrometry techniques. *Journal of Forensic Sciences* **29** (3) 714–722.

Keto, R.O. & Wineman, P.L. (1991), Detection of petroleum-based accelerants in fire debris by target compound gas chromatography/mass spectrometry. *Analytical Chemistry* **63** 1964–1971.

Kinard, W.D. & Midkiff, C.R. (1990), Arson evidence container evaluation II: KAPAK bags – a new generation. Presented: *42nd Annual Meeting, American Academy of Forensic Sciences, Cincinnati, OH* February, 1990.

King, R.W. (1988), Petroleum: its composition, analysis and processing. *Occupational Medicine: State of the Art Reviews* **3** (3) 409–430.

Kobus, H.J., Kirkbride, K.P. & Maehly, A. (1987), An adsorption sampling method combined with capillary column gas chromatography and cryogenic focussing for trace analysis of volatile organic compounds. *Journal of the Forensic Science Society* **27** 307–314.

Krull, I.S., Swartz, M. & Xie, K.H. (1983), The use of multiple detection in the gas chromatographic analysis of organic nitro compounds and explosives (GC-ECD/PID). *Proceedings of the International Symposium on the Analysis and Detection of Explosives* March 29–31, US Department of Justice, Federal Bureau of Investigation, 107–122.

Kubler, D.G. & Stackhouse, C. (1982), Relative hydrocarbon detectability by flame ionization detection for various isolation methods. *Arson Analysis Newsletter* **6** (4) 73–82.

Lafleur, A.L. & Mills, K.M. (1981), Trace level determination of selected nitroaromatic compounds by gas chromatography with pyrolysis/chemiluminescent detection. *Analytical Chemistry* **53** (8) 1202–1205.

Lafleur, A.L., Morriseau, B.D. & Fine, D.H. (1978), Explosives identification in post-blast residues and other matrices using gas chromatography (gc) and high performance liquid chromatography (hplc) combined with a NO-specific detector. *Proceedings: New concepts symposium and workshop on detection and identification of explosives*, October 30–November 1, 1978, 597–598, NTIS: Springfield, VA.

Langhorst, M.L. (1981), Photoionization detector sensitivity of organic compounds. *Journal of Chromatographic Science* **19** 98–103.

Lentini, J.J., Tontarski, R.E., DeHaan, J.D., O'Donnell, J.F. & Rogers, B.J. (1989), Glossary of terms related to chemical and instrumental analysis of fire debris. *Fire and Arson Investigator* **40** (2) 25–34.

Lentini, J.J., Fultz, M.L., Armstrong, A., Davie, B., DeHaan, J., Henderson, R., O'Donnell, J., Rogers, B.J. & Small, J. (1990), Forensic Science Committee Position on comparison samples. *Fire and Arson Investigator* **41** (2) 50–51.

Leung, K. & Yip, H.L. (1970), Identification of light-petroleum products by gas chromatography. *The Canadian Society of Forensic Science Journal* **3** 42–51.

Lipska, A. & Wodley, F. (1969), Isothermal pyrolysis of cellulose: kinetics and gas chromatographic mass spectrometric analysis of the degradation products. *Journal of Applied Polymer Science* **13**, 851–865.

Lloyd, J.B.F. (1982), Capillary column gas chromatography in the examination of high relative molecular mass petroleum products. *Journal of the Forensic Science Society* **22** 283–287.

Lloyd, J.B.F. (1991), Forensic explosives and firearms traces: trapping of HPLC peaks for gas chromatography. *Journal of Energetic Materials* **9** 1–17.

Long, W.C. (1978), Lab exercise number nine, identification of dyes in gasoline. *Arson Analysis Newsletter* **2** (3) 2–4.

Loscalzo, P., DeForest, Pl. & Chao, J.M. (1977), Results of arson detection survey. *Arson Analysis Newsletter* **1** (6) 4–8.

Lucas, D.M. (1960), The identification of petroleum products in forensic science by gas chromatography. *Journal of Forensic Sciences* **5** (2) 236–247.

Mach, M.H. (1977), Gas chromatography-mass spectrometry of simulated arson residue using gasoline as an accelerant. *Journal of Forensic Sciences* **22** (2) 348–357.

Mach, M.H., Pallos, A. and Jones, P.F. (1978a), Feasibility of gunshot residue detection via its organic constituents. Part I: analysis of smokeless powders by combined gas chromatography-chemical ionization mass spectrometry. *Journal of Forensic Sciences* **23** 433–445.

Mach, M.H., Pallos, A. & Jones, P.F. (1978b) Feasibility of gunshot residue detection via its organic constituents. Part II: a gas chromatography-mass spectrometry method. *Journal of Forensic Sciences* **23** 446–455.

Macoun, J.M. (1952), The detection and determination of small amounts of inflammable hydrocarbons in combustible materials. *Analyst* **77** 381.

Martz, R.M., Munson, T.O. & Laswell, L.D. (1983) Identification of smokeless powders and their residues by capillary column gas chromatography/mass spectrometry. *Proceedings of the International Symposium on the Analysis and Detection of Explosives* March 29–31, US Department of Justice, Federal Bureau of Investigation, 245–254.

Mann, D.C. (1987), Comparison of automotive gasolines using capillary chromatography I: comparison methodology. *Journal of Forensic Sciences* **32** (3) 606–615.

Mann, D.C. (1987), Comparison of automotive gasolines using capillary gas chromatography II: limitations of automotive gasolines in casework. *Journal of Forensic Sciences* **32** (3) 616–628.

Mann, D.C. (1990), Gasoline residue: evaporation or combustion? Presented: 42nd Annual Meeting. American Academy of Forensic Sciences, Cincinnati, OH. February, 1990.

Mann, D.C. & Gresham, W.R. (1990), Microbial degradation of gasoline in soil.

Journal of Forensic Science **35** (4) 913–923.
McNair, H.H., Ogden, M.W. & Hensley, J.L. (1985), Recent advances in gas chromatography. *American Laboratory* **17** (8) 15–16, 18–19.
Midkiff, C.R. (1975), Brand identification and comparison of petroleum products – a complex problem. *Fire and Arson Investigator* **26** (2) 18–21.
Midkiff, C.R. (1978), Liquids in arson evidence. *Arson Analysis Newsletter* **2** (6) 8–20.
Midkiff, C.R. (1980), Applications of polar liquid phases to the gas chromatographic examination of arson evidence. *Arson Analysis Newsletter* **3** (6) 1–22.
Midkiff, C.R. (1982), Arson and explosive investigation. In *Forensic Science Handbook*, Richard Saferstein, ed., Prentice-Hall, Inc. 223–266.
Midkiff, C.R. & Washington, W.D. (1972), Gas chromatographic determination of traces of accelerants in physical evidence. *Journal of the Association of Official Analytical Chemists* **55** (4) 840–845.
Moss, R.D., Guinther, C.A. & Thaman, R.N. (1982), The analysis of gasoline dye in fire debris samples by thin layer chromatography. *Arson Analysis Newsletter* **6** (1) 1–14.
Munder, A., Chesler, S.N. & Wise, S.A. (1990), Capillary supercritical fluid chromatography of explosives: investigations on the interactions between the analytes, the mobile phase and the stationary phase. *Journal of Chromatography* **521** 63–70.
Nowicki, J.F. (1981), Control samples in arson analysis. *Arson Analysis Newsletter* **5** (1).
Nowicki, J.F. (1990), An accelerant classification scheme based on analysis by gas chromatography/mass spectrometry (GC-MS). *Journal of Forensic Sciences* **35** (5) 1064–1086.
Nowicki, J.F. & Strock, C. (1983), Comparison of fire debris analysis techniques. *Arson Analysis Newsletter* **7** (5) 98–108.
Parker, B.P., Rajeswaran, P. & Kirk, P.L. (1962), Identification of fire accelerant by vapor phase chromatography. *Microchemical Journal* VI 31–36.
Pate, C.T. & Mach, M.H. (1978), Analysis of explosives using chemical ionization mass spectroscopy. *International Journal of Mass Spectrometry and Ion Physics* **26** 267–277.
Patterson, P.L. (1986), Recent advances in thermionic ionization detection for gas chromatography. *Journal of Chromatographic Science* **24** 41–52.
Patterson, P.L. (1990), Oxygenate fingerprints of gasolines. *DET Report* No. 18, October, 1990, 8–12.
Pearce, W.E. (1977), Study of gasoline dyes. *Arson Analysis Newsletter* **1** (3).
Penton, A. (1983), Determination of nitro explosives by GC utilizing an on-column capillary injector. *Proceedings of the International Symposium on the Analysis and Detection of Explosives* March 29–31, US Department of Justice, Federal Bureau of Investigation, 123–128.
Petroleum (Refinery Process, Survey) (1982), Kirk/Othmer, *Encyclopedia of Chemical Technology* **17** Third Edition, John Wiley & Sons, Inc.
Phillips, J.H., Coraor, R.J. & Prescott, S.R. (1983), Determination of nitroaromatics in biosludges with a gas chromatograph/thermal energy analyzer. *Analytical*

Chemistry **55**, 889–892.

Reeve, V., Jeffery, J., Weihs, D. & Jennings, W. (1986), Developments in arson analysis: a comparison of charcoal adsorption and direct headspace injection techniques using fused silica capillary gas chromatography. *Journal of Forensic Sciences* **31** (2) 479–488.

Rudolph, T.L. (1983), A scheme for the analysis of explosive residues. *Proceedings of the International Symposium on the Analysis and Detection of Explosives* March 29–31, US Department of Justice, Federal Bureau of Investigation, 71–78.

Rudolph, T.L. & Bender, E.C. (1983), *Proceedings of the International Symposium on the Analysis of Explosives*, 29 March 1983, US Department of Justice, FBI, pp. 71–89.

Russell, L.W. (1981), *J. Forensic Sciences* **21** 317–323.

Saferstein, R. & Park, S.A. (1982), *J. Forensic Sciences* **27** (3) 484–494.

Sanders, W.N. & Maynard, J.B. (1968), Capillary gas chromatographic method for determining the C_3–C_{12} hydrocarbon in full range motor gasolines. *Analytical Chemistry* **40** (3) 527–535.

Schwanebeck, W. (1984), A method for testing packaging materials and systems used for the storage of arson debris. Abstract No. 94, International Association of Forensic Sciences Meeting, Oxford, UK.

Smith, R.M. (1983), Mass chromatographic analysis of arson accelerants. *Journal of Forensic Sciences* **28** (2) 318–329.

Smith, R.M. (1987), Arson analysis by mass chromatography. *Forensic Mass Spectroscopy*, Jehuda Yinon (ed.), Boca Rotan, FL: CRC Press, Inc., 131–159.

Sullivan, J.J. (1977), Detectors in *Modern Practice of Gas Chromatography*, Robert L. Grob (ed.), John Wiley & Sons 228–254.

The Supelco Reporter (1988), For high resolution or high speed gc analyses, use 0.20 mm ID capillary columns. **7** (6) 1–3.

The Supelco Reporter (1989), A 100-meter capillary column simplifies many difficult petroleum analyses. **8** (2) 4–6.

Tamiri, T. & Zitrin, S. (1986), Capillary column gas chromatography/mass spectrometry of explosives. *Journal of Energetic Materials* **4** 215–237.

Thornton, J.I. & Fukayama, B. (1979), The Implications of refining operations to the characterization and analysis of arson accelerants. Part I. Physical Separation. *Arson Analysis Newsletter* May 1–16 Part II. Chemical conversions, Treating Processes, and Subsidiary Processes. August 1–16.

Tontarski, R.E. (1983), Evaluation of polyethylene containers used to collect evidence for accelerant detection. *Journal of Forensic Sciences* **28** (2) 440–445.

Tontarski, R.E. (1985), Using absorbents to collect hydrocarbon accelerants from concrete. *Journal of Forensic Sciences* **30** (4) 1230–1232.

Tontarski, R.E. & Strobel, R.A. (1982), Automated sampling and computer-assisted identification of hydrocarbon accelerants. *Journal of Forensic Sciences* **27** (3) 710–714.

Tranthim-Fryer, D.J. (1990), The application of a simple and inexpensive modified carbon wire adsorption/solvent extraction technique to the analysis of accelerants and volatile organic compounds in arson debris. *Journal of Forensic Sciences* **35**

(2) 271–280.

Trimpe, M.A. (1991), Turpentine in arson analysis. *Journal of Forensic Sciences* **36** (4) 1059–1073.

Twibell, J.D. & Home, J.M. (1977), Novel Methods for direct analysis of hydrocarbons in crime investigation and air pollution studies. *Nature* **268** (5622) 711–713.

Twibell, J.D., Home, J.M. & Smalldon, K.W. (1981), A splitless curie point pyrolysis capillary inlet system for use with the adsorption wire technique of vapour analysis. *Chromatographia* **14** (6) 366–370.

Willson, D. (1977), A unified scheme for the analysis of light petroleum products used as fire accelerants. *Forensic Science* **10** 243–252.

Willson, D. (1984), Automatic analysis and data processing as an aid to the routine analysis by gas chromatography (gc) of samples submitted by fire investigators. Abstract No. 83, *International Association of Forensic Sciences Meeting, Oxford, UK.*

Woychesin, S. & DeHaan, J. (1978), An evaluation of some distillation techniques. *Arson Analysis Newsletter* **2** (5) 1–16.

Yandler, M.W. & Hanley, W.P. (1978), Post explosion detection of nitrate esters. *Proceedings: New concepts symposium and workshop on detection and identification of explosives*, October 30–November 1, 1978, 607–609. NTIS: Springfield, VA.

Yinon, J. (1977), Analysis of explosives. *CRC Critical Reviews in Analytical Chemistry*, Cleveland, Ohio: CRC Press, **7** (1) 1–35.

Yinon, J. (1982), Mass spectrometry of explosives: nitro compounds, nitrate esters, and nitramines. *Mass Spectrometry Reviews*, George R. Waller (ed.), New York: John Wiley & Sons, **1** (3) 257–307.

Yinon, J. (1987), Mass spectrometry of explosives. *Forensic Mass Spectrometry*, Jehuda Yinon (ed.), Boca Raton, FL: CRC Press, Inc., 106–130.

Yip, I.H.L. (1982), A sensitive gas chromatographic method for analysis of explosive vapours *The Canadian Society of Forensic Sciences Journal* **15** (2) 87–95.

Yip, I.H.L. & Clair, E.G. (1976), A rapid analysis of accelerants in fire debris. *The Canadian Society of Forensic Science Journal* **9** (2) 75.

Zitrin, S. (1986), Post explosion analysis of explosives by mass spectrometric methods. *Journal of Energetic Materials* **4** 199–214.

6

Pyrolysis gas chromatography in forensic science

Peter R. DeForest, D. Crim.
John Jay College, New York, NY, USA

Ian R. Tebbett, Ph.D.
Department of Pharmacodynamics, University of Illinois at Chicago, Chicago, IL, USA

A. Karl Larsen, B.A.
Illinois State Police, Suburban Chicago Laboratory, Maywood, IL, USA

6.1 INTRODUCTION

One activity which distinguishes criminalistics from other related scientific fields such as analytical chemistry is the need to attain the highest degrees of discrimination between very similar samples in attempts to approach the goal of individualization. This goal is desirable in casework dealing with evidence which is used to demonstrate a link between a suspect and a victim or a suspect and a scene. Such evidence is often referred to as associative evidence. To truly attain individualization with this type of evidence it would be necessary to show that the two items being compared are more similar to each other than they are to any other like objects in the universe. Stated somewhat differently, the comparison should be good enough, in the context of the variation within the population of like items, to allow the inference to be drawn that the two items share a unique common origin.

At the present time this degree of refinement is attained only with limited types of evidence such as comparisons involving detailed physical patterns, physical markings or physical matches. Examples of these physical pattern evidence types would include the microstriae in rifling marks on fired bullets, friction ridge detail in fingerprints, random or accidental acquired characteristics on footwear outsoles and the mating of random fracture surfaces. If one sets aside the special case of the chemically coded sequence information in the form of DNA in the genome, individualization cannot be achieved on the basis of chemical composition. The requisite degree of discrimination needed for individualization based on chemical composition cannot be attained or even closely approached with currently available techniques which offer far less discrimination than is required for true individualization. Here it is necessary to use statements to the effect that the two samples 'could have' had a common origin. Such conclusions are based on the failure of techniques which are known to offer high degrees of discrimination to distinguish between the samples. Although there are other factors to be taken into account when two samples are found to be indistinguishable, the better the potential discrimination of the selected technique which fails to reveal significant differences, the better the likelihood of a common origin.

In addition to the need for high discrimination, techniques used in comparing associative evidence must often satisfy two additional stringent criteria. Forensic samples encountered in casework may be extremely small and may be chemically intractable. Further, it is often desirable to use only a portion of the sample, and useful approaches should be either nondestructive or so sensitive that only a small portion of a small sample is consumed in analysis. Thus applicable techniques used for this purpose must be very sensitive to be able to maximize the extraction of information from samples that are often quite small. They should also be capable of dealing with chemically intractable samples. For complex organic materials gas chromatography satisfies the first criterion and pyrolysis the second. Combining the two would appear to be ideal for individualization work.

6.2 GAS CHROMATOGRAPHY AND PYROLYSIS GAS CHROMATOGRAPHY

Gas chromatography (GC) was introduced into analytical chemistry in the early 1950s. It got off to a slow start. Those scientists who wished to utilize or explore the virtues of GC for particular analytical problems often built their own rudimentary instruments. By 1960 increasingly large numbers of papers describing gas chromatography applications were appearing in analytical chemistry journals. Also by this time several lines of commercially produced instruments were on the market. Some were made by start-up companies who began manufacture on a small scale in modest rental space or even basements and garages. Many of these start-up companies grew to be serious competitors for more established instrumentation companies. Some, after establishing themselves and gaining name recognition for their product lines, were bought out and their product lines absorbed by larger companies.

GC is a very sensitive technique which is capable of dealing with small amounts of analyte contained within complex mixtures. It has the limitation that the species to be analyzed must be capable of being volatilized. In the case of volatile samples of forensic interest, it is directly applicable. Gas chromatography used alone or used with mass spectrometry has been applied to the analysis of blood alcohol, many drugs of abuse and debris from fire scenes suspected of containing accelerants. It is especially useful for complex mixtures of volatile organic substances. For such samples GC has no peer.

GC's requirement of volatile samples does limit its direct application to many other kinds of forensic evidence. Polymers and other types of macromolecular evidence are clearly nonvolatile and cannot be analyzed by GC directly. Many are also insoluble in a range of solvents and are thus quite intractable to many other analytical techniques. Means of breaking some of these samples down into simpler compounds to facilitate analysis by GC undoubtedly would have been attractive. At the time that GC was introduced the concept of deliberate degradation of samples prior to analysis was not new to analytical chemists. Commonly used means of preparatory sample degradation included hydrolysis, enzyme digestion, as well as pyrolysis. Prior to the advent of GC the analysis of pyrolyzates was rather crude. Wet chemical tests for specific evolved gases such as ammonia, carbon dioxide, etc. were common. The marrige of GC and pyrolysis to yield pyrolysis-GC (PyGC), once conceived of, had obvious possibilities. Thus the capability of analyzing very small, chemically intractable samples, with the potential for a high degree of discrimination, made it a very attractive tool for use with certain physical evidence problems. These three anticipated advantages were quickly realized, as a brief review of the history of PyGC will show (see below).

Pyrolysis has been defined as the transformation of a compound into another substance or substances through the agency of heat alone (Hurd 1929). Generally pyrolysis involves the breakdown of the substance under examination into smaller compounds, as a result of thermal decompositions or degradations. Pyrolysis can however also include the formation of polymers and complex molecules under the influence of heat.

Pyrolysis has been used successfully in a one- or two-stage combination of pyrolysis with mass spectrometry (MS) (Bradt *et al.* 1953, Happ & Maier 1964, Voigt & Fischer 1964), infra red spectrophotometry (IR) (Brown *et al.* 1963, Cleverley & Herrmann 1960, Guischon & Henniker 1964, Harms 1953, Kruse & Wallace 1953, Kupfer 1962), ultraviolet spectrophotometry (UV) (Silverman 1958), gas chromatography (GC) (Voigt & Fischer 1963, Cleverley & Herrmann 1960, Guiochon 1964, Kupfer 1962, Silverman 1958, Andrew 1963, Bhatnager & Dhont 1962, Bombaugh *et al.* 1963, Borer *et al.* 1960, Brauer 1965, Ettre & Varadi 1962, Fontan *et al.* 1964, Glassner & Pierce 1965, Jones & Moyles 1961, Karr *et al.* 1963a,b, Keulemans 1958, Stanley & Peterson 1962). and even spot tests (Fiegl & Jungreis 1958). Pyrolysis is ideally suited for combination with gas chromatography. Different pyrolysis devices can be directly connected to the inlet of the gas chromatograph and a suitable gas chromatographic system can resolve the compounds produced by pyrolysis into a highly specific pattern of peaks. This pattern, referred to as a pyrogram, can be used as a 'fingerprint' in the identification and comparison of complex materials.

6.3 HISTORICAL DEVELOPMENT OF PYROLYSIS GAS CHROMATOGRAPHY

As noted above, pyrolysis as an analytical technique predates the development of chemical instrumentation and the subsequent displacement of many wet chemical techniques. For many years chemists would analyze certain difficult samples by heating them in a vessel which excluded air until the desired degree of decomposition had taken place. Simple tests might be performed on some of the volatiles evolved (presence of ammonia, etc.) or the tar-like residue remaining in the vessel might be extracted and analyzed. It is not surprising then that chemists would apply instrumental techniques to the analysis of pyrolyzates as these techniques became available.

Zemany (1952) was the first to suggest the use of pyrolysis combined with instrumental analysis of the resulting mixture as a general approach to the identification of organic materials. He argued that, like the ions in a mass spectrum, the identity and relative amounts of the various products from a given material should always be constant. Zemany also suggested that mass spectrometry, infra-red spectroscopy or any other method of examining the pyrolysis products could be employed to identify the original material. He analyzed pyrolyzates of polymers by mass spectrometry. This was apparently the first use of an instrumental method to analyze a pyrolyzate.

When gas chromatography was in its infancy and commercial instruments were as yet unavailable, experiments with PyGC began. At first a batch processing mode was used. The sample was pyrolyzed in a separate vessel and later the volatile pyrolysis products were transferred to the GC inlet.

Davidson *et al.* (1954) were the first to apply gas chromatography to the analysis of pyrolyzates for identification purposes. As Zemany had suggested, their work emphasized that the pattern of peaks in a pyrogram provided a characteristic fingerprint which could be used as a means of identification of the original substance. Subsequently, Burns *et al.* (1957) reported the use of PyGC as a new tool for polymer identification.

Keulemans (1958) emphasized Zemany's theories by suggesting that pyrolysis and gas chromatographic separation of the resulting complex was analogous to the principles of mass spectrometry, that is, fragmentation and magnetic separation. De Angelis *et al.* (1958) made a step forward by studying the pyrolyzates of a number of polymers on three different columns.

It was soon realized that GC had unique attributes that would make pyrolysis of microgram sized samples in the carrier gas stream desirable (Martin 1959). This became known as on line pyrolysis GC. The heretofore unrivaled separating power of GC combined with the advent of ionization detectors resulted in the birth of a formidable combination. The cost of the instruments was surprisingly modest, even for the early 1960s. A Wilkens Aerograph (later absorbed by Varian) HyFI, hydrogen flame ionization, GC sold for about $1000. Forensic scientists were among the first to recognize the advantages that 'on-line' pyrolysis-GC would have for the analysis and comparison of certain difficult samples of forensic interest (Kirk 1963, Raddell & Strutz 1959).

Martin (1959) was the first to report the application of one-stage pyrolysis gas chromatography to the analysis of noncondensable bases produced in rapid degradative studies of solids. He emphasized the essential feature of carrying out the reaction directly in the stream of the chromatography carrier gas.

Jones & Moyles (1961) established the sensitivity of pyrolysis gas chromatography when they demonstrated the advantage of carrying out pyrolysis using microgram quantities of material rather than milligrams. Using a smaller sample size also minimized the incidence of secondary reactions occurring in the pyrolysis chamber. Similar work on optimizing sample size for PyGC was performed by Simon & Giacobbo (1965) who specifically studied the effect of film thickness on secondary reactions and on rates of decomposition. A sampling method was also developed for powders. This technique allowed the pyrolysis of 10–50 μg of material in the form of a thin homogeneous powder layer applied to the surface of the heating filament, and was particularly useful for forensic work involving the analysis of microgram quantities of insoluble materials (Levy 1963, Levy *et al.* 1964, Levy & Gesser 1965).

Attempts at standardization of methodologies were made by a number of workers who incorporated thermocouples into filament pyrolysis units in order to estimate the pyrolysis temperature (Voigt & Fischer 1964, Garzo & Szekely 1964, Voigt 1964). Voigt & Fischer (1964) developed a procedure by which the thermocouple reading on the pyrogram was recorded via an automatic switching device. Esposito (1964) demonstrated the applicability of adding an internal standard to the sample for quantitative pyrolysis gas chromatography. In further modifications, Parson (1964) utilized a gas density balance to determine the molecular weights of the pyrolysis products.

The quality of the results from PyGC studies is greatly dependent on the performance of the gas chromatographic system. The need to reveal a larger portion of the pyrolyzate spectrum with greater detail has brought the application of more advanced methods of gas chromatography to PyGC studies.

Stanley & Peterson (1962) were the first to connect a pyrolysis unit to a capillary gas chromatographic column. This work was carried out under isothermal conditions and, as would be expected, resulted in much greater resolution of the pyrolysis products than the packed columns. Excellent results were later obtained by combining pyrolysis with temperature programmed gas chromatographs (Braver 1965, Simon & Giacobbo 1965). Subtle changes in the heating rates of the GC oven could be used to effect greater separation of the most relevant parts of the pyrogram. The advantages of using cryogenic temperature programming in PyGC studies were also demonstrated (Barbour 1965).

6.4 METHODS FOR COMBINING THE PYROLYSIS PROCESS WITH GAS CHROMATOGRAPHY

6.4.1 Theoretical considerations

There are two general types of use of PyGC with respect to forensic needs. One involves identification, the other individualization. The latter is the most demanding. Identification of many polymers is easily accomplished with PyGC. PyGC is a particularly powerful tool for this purpose. It makes determining the generic classes of paint vehicles, fiber polymers and many other evidence polymers almost trivial. Some alternative analytical methodologies such as FTIR microscopy can also accomplish generic identifications and have the advantage of being nondestructive. Where mere identifications are called for in situations where the amount of sample available is limited, such alternatives should be carefully considered. However, PyGC is likely to be the method of choice when the need is the detection of subtle differences within a single class of materials, that is, one wishes to approach individualization.

As might be expected from our earlier discussion, individualization presents a far more challenging problem than is the situation where only identification is necessary. The potentials of PyGC for the individualization problem in forensic science casework have been reviewed (DeForest 1969, 1974). The central challenge can be expressed as the need to detect minor variations among samples which are generally similar with respect to their major components. This translates into the requirement of being able to find subtle structural or trace compositional variations within a matrix that is overwhelmingly similar among closely similar samples of the same class. This is not a trivial problem. It places acute demands on the PyGC system. The chromatographic separation must be optimized so that minuscule minor peaks arising from trace differences are not lost among the larger peaks produced by the constituents common to all of the samples of the class.

It is a relatively simple operation to connect a pyrolysis unit and a gas chromatograph (Janak 1960a,b, Levy & Gesser 1964, 1965, Nelson & Kirk 1962). Two methods have been used to facilitate this combination: by connecting the pyrolysis unit directly into the inlet of the gas chromatograph or via a suitable valve, in which case the pyrolysis unit has an independent flow path.

When a pyrolysis unit is directly connected to the inlet of the gas chromatograph a number of requirements must be met for the successful gas chromatographic separation of the pyrolysis products. A number of variables such as duration of the pyrolysis process, the flow rate through the unit, the pressure, the quantity of pyrolyzed material, the volume of the pyrolysis chamber, and the nature of the carrier gas, must be carefully controlled in order to optimize the gas chromatographic separation.

Introduction of the pyrolysis products into the carrier gas stream as a single well defined plug is essential if good chromatographic resolution is to be achieved. Dandoy (1962) suggested that pyrolysis durations of 10–15 seconds should be employed in order to achieve this, whereas Voigt (1964) reported 30 seconds as being the maximum limit. Pyrolyses have been carried out by various workers for anywhere between 1 and 30 seconds (Levy 1963, Lehman & Braver 1961). In actual fact, the maximum permissible duration of pyrolysis at which the resolution is still not affected varies, depending on several parameters of the gas chromatographic system. These include: the efficiency of the column, the total volume of the inlet system, carrier gas flow rate etc. Pyrolysis times can be considerably increased when temperature programming of the column is used. Microgram quantities of material can be completely pyrolyzed within a very short time (0.3–2 seconds). For pyrolysis of larger quantities of material 5–30 seconds may be required. Loss of resolution due to overloading of the column may occur when larger quantities of material are pyrolyzed and the pyrolyzate fed directly onto the column. The largest quantity which can be pyrolyzed without causing loss of resolution due to overloading is dependent on the capacity of the column. Capillary columns can handle only microgram quantities of material when the pyrolyzate is fed directly into the column. Obviously, the smallest quantity of material which can be studied by PyGC is limited by the sensitivity of the detection system used.

The pressure and the flow conditions in the inlet system of the gas chromatograph, that is, in the pyrolysis unit, are usually set to achieve optimum performance of the column. The permissible flow rates are therefore limited and will depend mainly on the diameter of the column and column efficiency. In order to obtain higher flow rates through the pyrolysis chamber without applying high pressures, Martin (1959) found it necessary to use a larger diameter column. A specially designed flow system which uses two sources of carrier gas and which permits selection of a wider range of pyrolysis chamber flows and pressures was described by Levy (1963).

When a pyrolysis unit is connected to the gas chromatograph via a valve the pyrolysis process can be carried out almost independently of the fact that it is followed by a gas chromatograph. The pyrolysis unit is first flushed with carrier gas and then closed; the pyrolysis is then carried out statically in the carrier gas. The valve is then switched to allow the carrier gas to take the pyrolysis products onto the column.

6.4.2 Types of pyrolysis unit

Three general types of pyrolysis unit have been used. These are as follows: Curie point units, filament units and oven or furnace type units. In Curie point pyrolyzers the sample is placed on or in contact with a small piece of ferromagnetic material, often in the form of a wire. Application of the sample onto the filament surface in the form of a thin film can be achieved by bringing the filament in contact with a solution of the sample followed by evaporation of the solvent. Sample coating of the filament is convenient for the uniform application of microgram or sub-microgram quantities of material. The coating method has numerous advantages, principally allowing the instantaneous pyrolysis of the sample. It is however largely limited to soluble samples. In addition, the structure of the film may differ from the structure of the solid sample and therefore the pyrolysis of the film may not be representative of the solid sample. Direct sampling of solid samples can be achieved by placing the sample into a removable sample container.

The wire containing the sample is inserted into the sample chamber. The sample chamber which is positioned within a radio frequency (RF) coil is purged with carrier gas to remove air. Pyrolysis is initiated by sending radio frequency energy to the coil. This radio frequency energy induces heating in the ferromagnetic sample wire. The wire is heated very rapidly up to its Curie point where it loses its ferromagnetic properties. Heating stops abruptly. The temperature at which the ferromagnetic properties are lost, the Curie point, varies with the composition of the material. Different ferromagnetic alloys with a range of Curie points are available to give the analyst a range of pyrolysis temperatures. The major advantages of Curie point pyrolyzers are the rapid heating of the sample carrier and the precision at which the final temperature is specified. Drawbacks are related to the possible reactivity or catalytic activity of the ferromagnetic alloy surface and to the difficulty of obtaining intimate contact between solid samples and the Curie point wire. Some analysts will clamp the sample in a bend in the wire to hold it in place. It should be clear that the contact between the sample and the wire is less than ideal under these circumstances.

With modern feedback controlled power supplies, electrically heated filament type pyrolyzers can achieve high heating rates and reasonably accurate final temperatures. Two types of electrically heated filament are in common use. They are frequently made of platinum to provide a non-reactive surface. The most common is the helical coil design. The other is the ribbon filament to aid in holding samples. The filaments are commonly affixed to the end of the probes which are inserted into an interface or alternatively directly into the inlet of the gas chromatograph. The inlet may be specially modified for this purpose. With the helical coil design intractable solid samples are often placed within a quartz or silica capillary tube which is held within the helical coil. The use of such a sample tube makes sample handling easier and may result in fewer problems with extreme temperature gradients within the sample. However, this convenience is obtained at the expense of much slower sample heating rates. In addition, if the capillary extends beyond the confines of the coil, as is often the case, condensation may take place within the cooler ends of the capillary tube.

The walls of the pyrolysis chamber (inlet or interface) are generally heated to minimize condensation of pyrolyzates. In most designs the probe containing the filament and the sample is inserted into this heated zone during purging to remove air from the system. It should be noted that this low level heating of the sample prior to pyrolysis may alter it. Certainly, volatiles dissolved in the sample may be expelled and flushed away by the purging stream of carrier gas. There may be some air remaining in the system when the heating begins. Thus the pre-pyrolysis heating of the sample may also alter certain samples by oxidizing them or by altering the degree of crosslinking within the sample. Some probe designs have allowed the sample to be purged in a cool zone prior to the time it is inserted into the heated zone and pyrolyzed. It should be clear that, in any case, the time the sample spends in the heated zone prior to pyrolysis should be held constant in a given series of experiments.

The Curie point filament units are the most commonly used types today. These, however, may not be the best for work with typical samples encountered in the forensic science laboratory. The oven or micro-furnace type pyrolyzer, although it has well recognized limitations, does have certain advantages with such samples. Unfortunately, some of the better designs of these are no longer in production. In one such micro-furnace design (Hamilton Multipurpose Sampling System) the sample was rapidly introduced into the preheated furnace zone after purging. The rapid sample introduction was achieved by altering the orientation of the tubular furnace from horizontal to vertical, allowing the sample to fall rapidly from the cool purge zone to a predetermined position, of known temperature, within the furnace. Other sample introduction methods with micro-furnace pyrolyzers have included the use of sample boats which could be moved from zone to zone by use of a magnet on the exterior of the sealed pyrolysis chamber. The advantages of the furnace units include uniform heating of the sample radially from the outside in, rather than from one side as with Curie point and some filament units.

In considering which types of pyrolysis units are most applicable to forensic samples, it is not necessarily true that what is best from a theoretical point of view is best in forensic applications. For example, there seems to be little question that the rapid heating of a thin film of sample coated onto the heating element of a flash type pyrolyzer offers the best prospects for obtaining uniform, reproducible and characteristic pyrolyzates. However, uniformly coating of a filament or Curie point wire can be impossible with intractable forensic samples. Many are simply not soluble. For this reason, it can be argued that the advantages that flash pyrolysis offers with solution coatable samples do not extend to chemically intractable forensic samples. It is apparent that all of the pyrolyzer designs have drawbacks with respect to many commonly encountered forensic samples. With chemically intractable samples of irregular shape, furnace units may offer the best prospects for uniform, reproducible heating. This question has not been adequately addressed experimentally.

6.4.3 Precision and accuracy in PyGC

With any analytical technique, but particularly with those techniques used for the comparison of forensic samples, the analyst must be concerned with the precision and accuracy of the procedure. The degree of precision and accuracy obtainable in PyGC studies has been the subject of considerable debate and attention (Bombaugh *et al.* 1963a, Barbour 1965, Barrall *et al.* 1963, Dhont 1964).

6.4.4 Qualitative reproducibility

When we consider the fact that pyrolysis is usually a complex system of reactions occurring simultaneously and consecutively, the reproducibility observed in PyGC is considered as being surprisingly good. Nelson & Kirk (1962) found that the pyrolyzates of different barbiturates produced a series of peaks which were reproducible in retention time and relative size as long as the column temperature and flow rate were kept constant, even though sample sizes were not accurately weighed. The qualitative reproducibility of pyrograms produced in two different laboratories using different pyrolysis units was demonstrated by Andrew *et al.* (1963). However the conditions of pyrolysis and sample preparation procedures must be exactly reproduced in order to prevent loss of precision.

6.4.5 Quantitative reproducibility

The quantitative reproducibility of PyGC has been studied and discussed by various workers (Bombaugh *et al.* 1963, Barbour 1965, Barrall *et al.* 1963a, Dhont 1964). Generally the deviations from the mean relative area of peaks in the pyrogram were determined following replicate analyses. Standard deviations have been reported as being between 0.5 and 3% which is comparable to other analytical techniques.

6.4.6 Accuracy

In most of the present applications of PyGC, particularly in forensic science, PyGC is used qualitatively for the comparison of two or more samples. Janak (1960b) concluded that the quantitative reproducibility of the method was not of the same order as other analytical methods and that PyGC should be considered only as semi-quantitative. However, the literature does contain examples of PyGC being used quantitatively with similar results to other procedures (Levy 1963, Arnold 1974, Barlow *et al.* 1961). The factors that affect the quantitative reproducibility of PyGC depend largely on the quantities of material involved and the principle of pyrolysis utilized. In each case optimum conditions must be sought to ensure reproducibility. Depending on the mechanism of their formation, different fragments appearing as peaks in the pyrograms would be affected to a different extent by variations in the experimental conditions. Levy observed that the relative quantities of the light hydrocarbons (methane, ethane, etc.) formed in pyrolysis of certain porphyrins are more sensitive to fluctuations in the experimental conditions than the relative amounts of heavier molecular weight fragments (Levy 1963). Most of the factors which affect the quantitative reproducibility depend on the overall set of conditions. Therefore the pyrolysis conditions should be chosen to minimize the effect of the factors which cannot be accurately controlled and are bound to fluctuate.

6.5 APPLICATIONS OF PyGC

It can be generalized that PyGC is applicable to any material if the pyrolyzate or a portion of it can be separated by gas chromatography. An extremely broad spectrum of compounds and materials satisfy this basic condition. The literature shows that PyGC has already been successfully applied to a wide spectrum of compounds and materials such as volatiles (Keulemans 1958, Dhont 1961, 1963b), non-volatiles (Andrew 1963, Fontan *et al.* 1964, Barbour 1965), organic synthetic polymers (Barbour 1965, DeForest 1974, Barrall *et al.* 1963a,b, Barlow *et al.* 1962, Lai & Locke 1983) and complex mixtures.

6.5.1 PyGC and forensic science

Members of Paul L. Kirk's group at the University of California at Berkeley were the first forensic science researchers to investigate the potential of PyGC with physical evidence problems (Fontan *et al.* 1964, Kirk 1963). They evaluated the application of PyGC to the identification of drugs and the individualization of paint and other polymers. Groten (1964) extended the analysis of polymers. This study was threefold, addressing the qualitative, quantitative and microstructural facets of pyrolysis gas-chromatography.

PyGC is now used routinely in forensic science laboratories for the comparison of adhesives (Curry 1987, Wright 1987), fibers (Bortnizk *et al.* 1971, Almer 1991), paints (McMinn *et al.* 1985, May *et al.* 1973, Audette & Percy 1978), polymers (Raddell & Strutz 1959, Simon & Giacobbo 1965, Hawley-Fedder *et al.* 1984), rubber compounds (Duncan 1988), tars etc. (Newlon & Booker 1979, Challinor 1983), see Figs. 6.1–6.3. Additional forensic science research with PyGC will be discussed in later sections. Many more applications for PyGC have been found in the last three decades. It is used in petroleum chemistry (Colling *et al.* 1986), bacteriology, art conservation (Chiarari *et al.* 1991), archeology, polymer chemistry, fossil fuel research (Granada *et al.* 1991), and in biomass studies among others. There are international conferences on PyGC and a journal devoted to the subject. Forensic scientists remain major users of the technique.

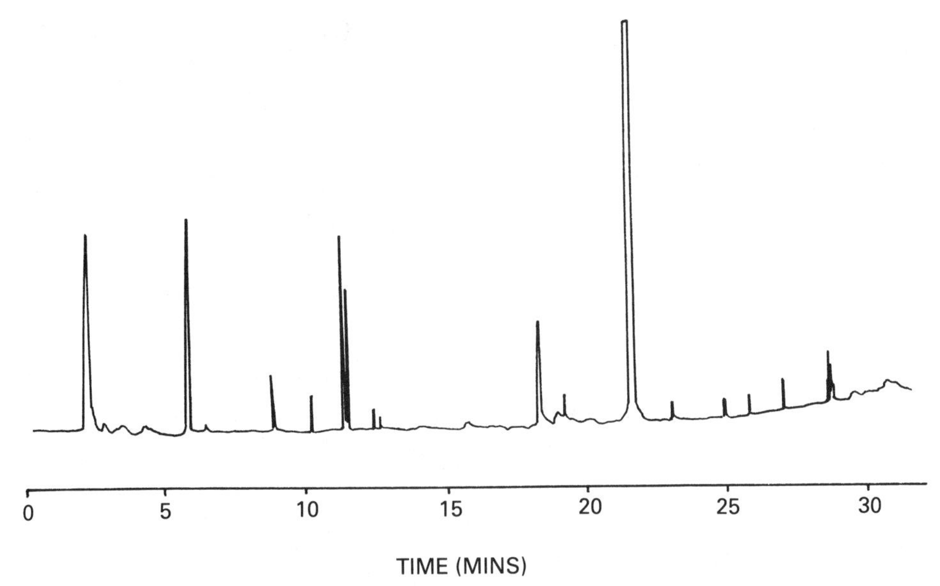

Fig. 6.1 Pyrogram of a nylon fiber using capillary gas chromatography.

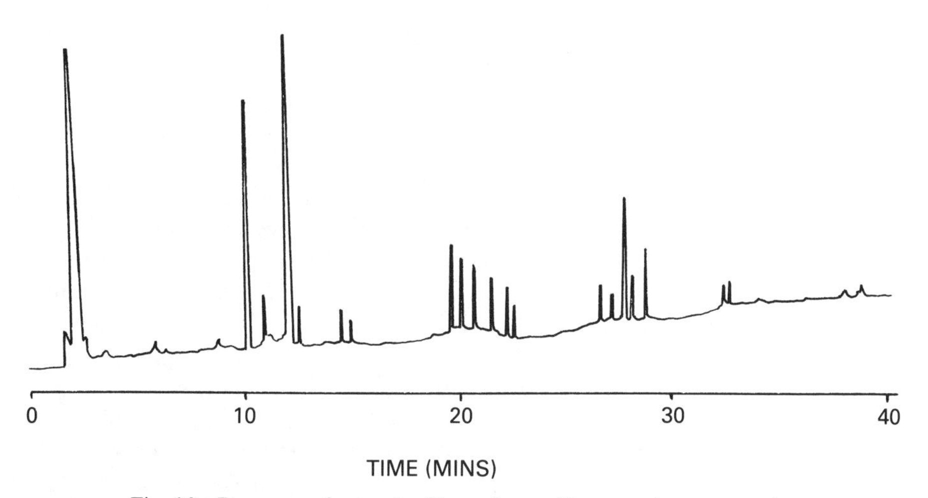

Fig. 6.2 Pyrogram of natural rubber using capillary gas chromatography.

6.5.2 Synthetic polymers

A milestone regarding progress in the forensic aspects of polymer analysis was documented by Bortniak *et al.* (1971). They were able to sub-characterize synthetic fibers within a given generic class. They validated the sensitivity of the technique by employing sample sizes in the microgram range. Further work focused on the microstructure of polymers, specifically on the composition of copolymers. May *et al.* (1973) reported a methodology that would facilitate data comparison of polymeric pyrolysis products, particularly plastics and paints, irrespective of laboratory origin.

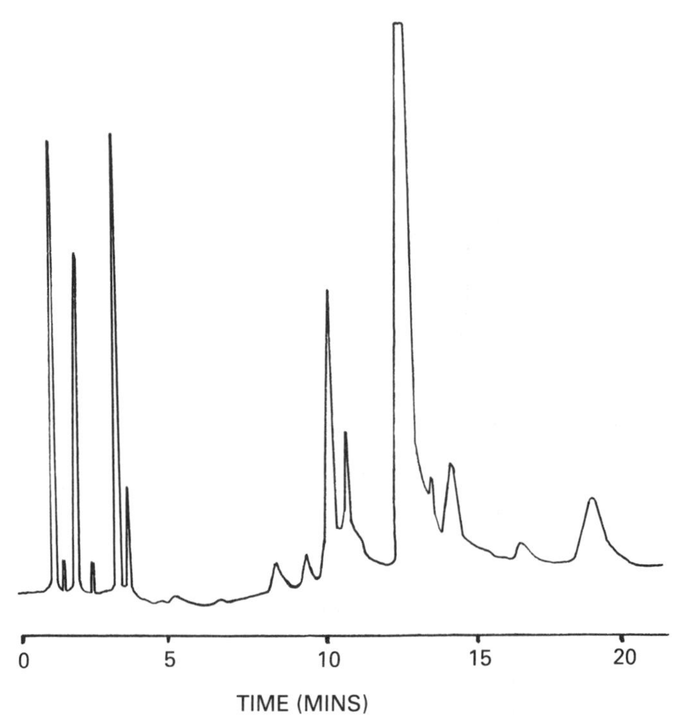

Fig. 6.3 Pyrogram of automobile paint.

A review of the potential individualization capabilities of PyGC applied to extremely large molecules inherent in many types of physical evidence was published in 1974 (DeForest). Part of the discussion was devoted to a description of the optimization of the instrumentation and conditions. Both Wheals & Noble (1974) and Stewart (1974) concentrated on the chemical components of automobile paints, specifically the organic binders and their differing formulations. In the former group, discrimination was accomplished for all but two manufacturers, and in the latter, all samples were distinguishable. This suggested a possible means of aging a motor vehicle. There is no end to the variety of evidence samples to which PyGC may be successfully applied. A comparative study of samples of chewing gum bases was conducted by Lloyd *et al.* (1974). Data were collected for both used and unused gums to be used as a reference set for a forensic case involving associative evidence. Material present in the throat of a murder victim was compared to material at the scene and to material found on the suspect. The pyrograms obtained showed that the materials were comparable. In addition, they corresponded to only one chewing gum in the reference collection. This was not true individualization but was helpful in solving the crime. Audette *et al.* (1978) developed a sequential technique that facilitated pyrolysis gas chromatography of micro paint samples which had previously been exposed to infra-red spectroscopy. They established the nondestructive nature of this prior analysis and its applicability to microchemical analysis. In the following year,

Newlon & Booker (1979) applied pyrolysis gas chromatography to the identification of smokeless powders. They were able to differentiate forty samples on the basis of their chemical composition.

Challinor (1983) reported a high degree of discrimination for disparate samples of forensic interest. His data were obtained with the use of capillary columns. The samples analyzed included paints, motor vehicle rubbers, fibers, adhesives and polyurethane foams. This work with capillary columns was extended by McMinn *et al.* (1985) and Almer (1991). Automobile paints were the focus of the former study and their identification was facilitated by combining mass spectral data with the chromatographic 'fingerprint'. This work illustrates the feasibility of a universal data base for reference purposes. The latter work was conducted on polyacrylonitrile fibers in order to sub-categorize this generic class. The results indicated that further classification, based on a variation in polymer composition, is possible.

6.5.3 Natural polymers

Work with natural products including biological macromolecules has nicely illustrated the power of PyGC. Early in the history of PyGC it was used to differentiate among bacterial stains. It was even used in the search for extraterrestrial life, and compact PyGC units were designed as onboard instrumentation on planetary probes. Extensive studies of fossil fuels such as coals and kerogens have relied heavily on PyGC and PyGCMS. Much of the material comprising these materials is of complex macromolecular structures and is difficult to study by other means of chemical analysis.

Unfortunately, the individualization problem has proven to be more difficult. Exploration of PyGCs use as a tool for the individualization of hair was reported in the late 1960s (DeForest 1969). It was not shown to be a viable technique for this purpose given the stage of technological development at the time. A similar study by Munson & Vick (1985) using newer technology further underscored the difficulty of the hair individualization problem. In 1979, the technique was used on bloodstains. Clausen & Rowe (1980) were able to distinguish fetal and adult hemoglobin based on pyrograms which reflected their subtle structural variations.

Research in improving and applying PyGC is continuing. We noted above that numbers of publications in PyGC applied to polymers appeared in 1991. That year also saw pyrolysis applied to the analysis of the organic constituents present in artistic and archeological objects as reported by Granada *et al.* (1991). This illustrates the diverse applications that this technique may offer.

6.6 PYROLYSIS GAS CHROMATOGRAPHY — STATE OF THE ART

For use of PyGC in individualization efforts as opposed to identification work, the highest attainable discrimination is desirable. This implies the use of state of the art techniques with respect to both the pyrolysis and the chromatography. With respect to chromatography, low bleed, high resolution columns are necessary. In general, the best would be very long, narrow bore, capillary columns. Packed columns can suffice for identification problems but are not well suited to individualization problems. The long, narrow bore, capillary columns will provide the best resolution. The bonded phase is important to minimize background signal due to column bleed. This is necessary if the smallest peaks are to be seen above the background. Low column bleed allows the use of higher sensitivity settings. To best resolve the large number of components expected in a typical pyrolyzate, temperature programming (or temperature gradient elution) is essential. This can be combined with cryofocusing to facilitate efficient transfer of the relatively large volume of the dilute pyrolyzate to a confined zone at the head of the column. Good chromatography requires that the sample be confined to as small a zone as possible before the chromatographic separation is initiated. On the other hand, the best conditions for pyrolysis require rapid sweeping of the pyrolysis products from the pyrolysis chamber as they are formed to minimize secondary reactions. This results in a relatively large volume of dilute pyrolyzate and seems to put the demands of the pyrolysis and those of the chromatography in conflict. A preconcentration technique such as cyrofocusing is the best way to resolve this problem (Rothchild & DeForest 1982). Sample splitting as a means of keeping the sample plug on the column small is objectionable from the point of view of wasting sample. With respect to most trace evidence problems, only a limited amount of sample will be available for analysis.

The need to detect peaks resulting in minor differences in otherwise similar samples requires that the most sensitive detectors available be used. Ionization detectors which have been carefully optimized will fill this need. For most work, detectors with a general response to a broad range of species, such as hydrogen flame ionization, will be best. However, for certain applications sensitive detectors which respond to limited classes of compounds may find use. This would be true if it is known that the variations seen among similar samples are mainly due to substances which give rise to pyrolysis products derived from the bulk of the sample if the intrasample variation was not reflected in these. In this hypothetical circumstance the selectivity of the detector would provide a sensitivity benefit with respect to detecting the products derived from the minor components.

6.7 THE FUTURE

Predicting the future has its risks. However, it is safe to predict that PyGC cannot remain static. Incremental refinements in PyGC will continue. We can also expect to see various spin-offs of PyGC into new analytical approaches with their own virtues. These will be of three forms. Some will be combinations of pyrolysis with analytical methods other than gas chromatography but with non-pyrolytic degradation or combine non-pyrolytic degradation with alternative analytical methodologies.

Pyrolysis mass spectrometry (PyMS) actually predates the development of PyGC. Despite this early start the potentials of the technique have not been fully exploited. Electron impact (EI) ionization of many types of pyrolyzates produce mass spectra that are unduly complex. This is because EI produces a number of fragments for each of the components resulting from the sample pyrolysis. Softer ionization techniques such as chemical ionization (CI) have not been thoroughly evaluated with respect to their potentials with pyrolyzates of interest in forensic science (Meuzelaar *et al.* 1984).

We cannot expect to degrade a sample without the attendant loss of information. Recognition that the necessity of this loss is part of a trade-off should be explicit. With chemically intractable samples the loss of information may be an acceptable price to pay in order to gain access to the otherwise inaccessible information locked within these difficult samples. It can be expected that modes of degradation may differ with respect to the quantity and the nature of the information lost. Exploration of modified pyrolytic and non-pyrolytic modes of degradation can be expected.

A study exploring the potentials of pyrolysis-supercritical fluid chromatography (PySFC) for forensic work revealed problems with the state of maturity of SFC instrumentation and raised questions about the mechanisms of thermal degradation. Nonetheless the use of supercritical fluid chromatography is expected to play a major role in the future examination of samples which cannot presently be examined by either GC or HPLC (Klesp *et al.* 1962, Meyers & Giddings 1966, Giddings *et al.* 1968, Gouw & Jentoft 1972, Gere 1983, Fjeldsted & Lee 1984, Martin & Boehm in press, Norris & Rawden 1984, Smith *et al.* 1984, Schwartz 1987, Peadon & Lee 1983, James & Martin 1952, Novotny 1985, Pentoney *et al.* 1986, West & Lee 1986, Jackson *et al.* 1986).

A supercritical fluid is a substance which exists in a state which has characteristics

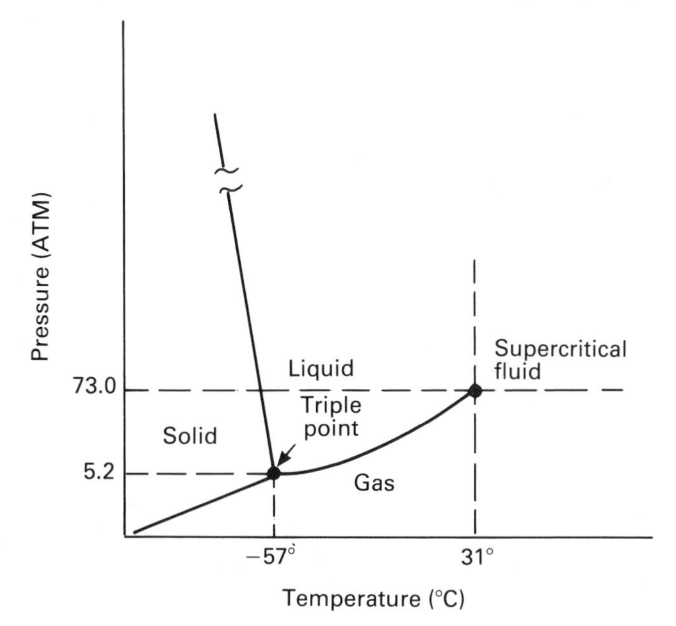

Fig. 6.4 Phase diagram of carbon dioxide.

of both a liquid and a gas. Fig. 6.4 shows a phase diagram for carbon dioxide. The values 73.0 atmospheres and 31°C are the critical pressure and temperature respectively. Normally, the critical temperature is much higher than the boiling point of the compound. At these conditions the liquid and gaseous phases of carbon dioxide have exactly the same densities. Although the theory of supercritical fluids has been discussed for many years, the technology needed for their use in routine analysis has only recently been introduced. Supercritical fluid chromatography was developed in an attempt to solve the analytical problems which neither GC nor HPLC could adequately handle. The density of a supercritical fluid (from 0.1 g/ml to 1.0 g/ml) is greater than that of a gas (0.001 g/ml), but less than a liquid (greater than 1.0 g/ml). It also has a viscosity range which falls between those of the carrier gas in gas chromatography and the mobile phase in high performance liquid chromatography (HPLC). The result of these two properties is that the supercritical fluid can take up more analyte than the carrier gas in GC. Although the diffusion coefficient of the supercritical fluid is almost 1000 times less than that of gas chromatographic carrier gas, it is 10 times greater than typical HPLC mobile phases. In theory, at least, greater resolution can therefore be achieved with supercritical fluid chromatography as compared to HPLC. In comparison with gas chromatography, SFC offers the advantage of greater solubility of the analyte, and allows the examination of those compounds which are not compatible with gas chromatography. Another benefit which SFC offers is the ability to use a wide range of detection systems. SFC can be coupled with the flame ionization detector and mass spectrometer as well as typical GC detectors. In addition, since the supercritical fluid has both the properties of a liquid and a gas, it is compatible with HPLC detectors such as an ultraviolet spectrophotometer. SFC can also be coupled to a Fourier transform infra-red spectrometer (FTIR), making this chromatographic technique almost universal in its methods of detection.

To date, supercritical fluid chromatography has been geared largely toward industrial and environmental applications. The food technology industry has made extensive use of SFC and supercritical fluid extraction (SFE). Oils and essences have been extracted efficiently and with less hazard to the analyst and environment when using this technique. The pharmaceutical industry has also used supercritical fluids in procedures for the characterization of products, including enantiomers, and trace contaminants. Presentations at the 4th International Symposium on Supercritical Fluid Chromatography and Extraction (1992) included extractions of fuels from soil, characterization of exhaust fumes and determination of explosive residues in water. Many of these have potential application to the forensic sciences.

Fig. 6.5 shows a schematic for a typical supercritical fluid chromatograph. The supercritical fluid extractor substitutes an extraction vessel for the column, and a collection trap for the detector. The manufacturers of these devices are continually improving the instruments, and automatic samplers with greater sample capacities are becoming available. Initially, HPLC columns were used for supercritical fluid chromatography, but these have now been improved to maximize efficiency of SFC separations. This drive to improve the instrumentation and accessories for SFE and SFC should not only continue, but accelerate as interest in this technique continues to grow.

The aspect of this technology which may generate the most interest in the forensic community is supercritical fluid extraction. The ability to extract meaningful information from a variety of matrices without the use of large volumes of solvents

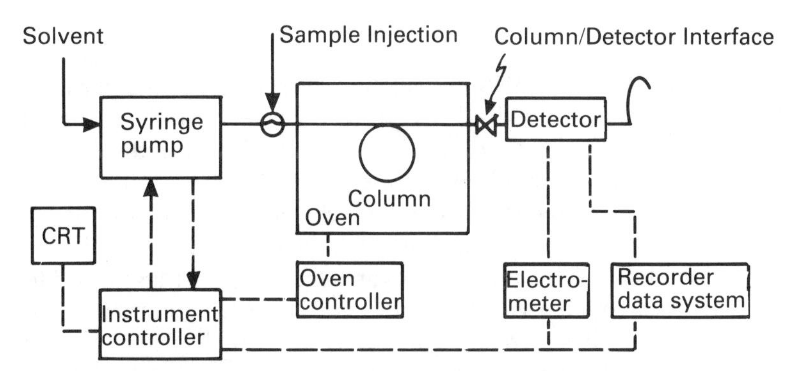

Fig. 6.5 Block diagram of a capillary supercritical fluid chromatograph.

is extremely attractive. The analyst should be able to make use of very small and complex samples and obtain extracts which are suitable for confirmatory identification via other instrumentation. In addition, like SFC, supercritical fluid extraction is compatible with a range of other instruments, allowing identification without added sample handling. With the ability to extract everything from drugs to fuels from matrices as varied as biological fluids to soils and polymers, supercritical fluids are expected to play a greater role in forensic science as applications are developed.

With regard to the compatibility of pyrolysis and supercritical fluid chromatography, it is reasonable to expect that the enhanced solvating ability of a supercritical carrier fluid will allow chromatography of larger, more characteristic, fragments resulting from less harsh pyrolysis temperatures. This is important since one of the major limitations of PyGC for individualization is the fact that pyrolysis temperatures necessary to produce fragments of low enough molecular weights to be handled by a gas chromatographic column destroy useful information.

It should be noted that the domain of low temperature degradation is in need of more exploration. Prolonged low temperature heating of some samples can make them more chemically intractable by producing more crosslinking. Lower temperatures alone may not be the whole answer. Alternative means of degradation need to be explored, keeping in mind the balance between loss of information and ease of analysis of products. Such alternate means will not be particularly useful unless they are selective. They will be most useful if they give the analyst control over the particular kinds of bonds to be broken. Examples might include hydropyrolysis or selective degradation with tunable lasers. Pyrolysis in the presence of specific reagents or catalysts are other alternatives for exerting control over the degradation.

In addition to new technologies, unexplored applications for PyGC still remain. These might include low temperature pyrolysis, or thermal desorption, of volatiles from selected biological stains. Stains which are recognizable by their distinctive odors when heated (urine, sweat, etc.) are the most obvious candidates for this approach. Other problem areas could include exploiting some of the information contained within the organic fraction of soil samples by pyrolyzing soils directly or by pyrolyzing extracts of them.

6.8 CONCLUSIONS

Attention must be given to the question of long term stability of analytical conditions in PyGC. In most cases problems of this type can be traced to the chromatography rather than the pyrolysis. This is also often the source of most of the difficulties with respect to limitations in the ability to use interlaboratory data. Most experience has shown that the pyrolysis conditions are much easier to standardize than the chromatographic ones. This is true at least with the current levels of discrimination power that are employed. Pyrolysis patterns are typically rather insensitive to modest changes in pyrolysis conditions. More important are the separation power of the column and its stability. The pyrolysis conditions *per se* are less important. However, it is generally true that high pyrolysis temperatures tend to yield fewer pyrolysis products but in greater quantity. These tend to be the same for similar samples. There is a loss of discrimination with extremely high pyrolysis temperatures. It is important to realize that the temperature experienced by major portions of the sample may be very different from that indicated on the instrument. Contact between the pyrolysis source (for example, filament) and the sample may be poor, and products produced at lower temperatures, due to both spatial and temporal gradients, may escape before they are exposed to the set point temperature. Advances in column technology such as bonded phase capillary columns have served to close the gap, but it is still generally true that the chromatographic conditions, including the means of transferring the pyrolyzate to the column, are more difficult to standardize than are the pyrolysis conditions. With such advances in chromatographic technology, it may be time to take a fresh look at the possibility of interlaboratory standardization of PyGC in forensic science.

PyGC is a versatile and powerful technique, but the results obtained with associative evidence must be interpreted cautiously. The authors have observed some analysts who apparently use PyGC because they are required to, or at least feel obligated to, use it with certain types of samples. Some may be concerned about questions being raised during court testimony. In any case some analysts appear to employ PyGC without giving much thought to questions of data interpretation. If the elution patterns of pyrolysis products are reasonably similar, this observation is used to support the opinion of a common origin for the samples being compared. Similarly, in situations where the patterns are different there is a tendency to dismiss the differences as being insignificant. Of course this is poor science. It is particularly dangerous because scientists can engage in this practice unknowingly. Vigilance is necessary. Criteria for conclusions of similarity or difference should be established ahead of time. It is clearly important to have experience with many samples of the general type being tested under the same experimental conditions. This provides the essential database for meaningful interpretation. If the particular PyGC method used doesn't have the *a priori* potential to refute a preliminary conclusion of common origin, it should not be used. This point should be obvious, but is often overlooked. The situation is not unique to PyGC. Unfortunately, the same criticism extends to many other types of comparative analyses used in forensic science laboratories. There is a clear need for increased numbers of more complete data bases in criminalistics.

The larger question of interpretation is deserving of prompt attention by forensic scientists. Should methods and techniques be utilized and relied on which have little *a priori* potential to provide significant additional information? Should techniques which accomplish little more than refined identifications be used as part of a series of tests where the ultimate question is one of common origin? What caveats are important for interpretations? Depending on the refinements employed, PyGC may provide information useful for the question of individualization or it may be merely an identification technique. It is the responsibility of the forensic scientist to know the essential difference.

REFERENCES

Almer, J. (1991), *J. Canadian Society of Forensic Science*, **24** 51.
Andrew, T.D., Phillips, C.S.G. & Semlyn, J.A. (1963), *J. Gas Chromatography*, **1** 27.
Arnold, P.E. (1964), *Proc. Soc. Anal. Chem.*, **1** 79.
Audette, R.J. & Percy, R.F.E. (1978), *J. Forensic Sciences*, **23** 672.
Barbour, W.M. (1965), *J. Gas Chromatography*, **3** 228.
Barlow, A., Lehrle, R.S. & Robb, J.C. (1961), *Polymer*, **2** 27.
Barlow, A., Lehrle, R.S. & Robb, J.C. (1962), *Makromol. Chem.*, **54** 230.
Barrall, E.M., Porter, R.S. & Johnson, J.F. (1963a), *J. Chromatography*, **11** 177.
Barrall, E.M., Porter, R.S. & Johnson, J.F. (1963b), *Anal. Chem.*, **35** 73.
Bhatnagar, V.M. & Dhont, J.H. (1962), *Nature*, **196** 769.
Bombaugh, K.J., Cook, C.E. & Clampitt, B.H. (1963), *Anal. Chem.*, **35** 1834.
Borer, K., Littlewood, B. & Phillips, C.S.C. (1960), *J. Norg. Nucl. Chem.*, **15** 316.
Bortniak, J.P., Brown, S.E. & Sild, E.H. (1971), *J. Forensic Sciences*, **16** 380.
Bradt, P., Dibeler, V.H. & Mohler, F.L. (1953), *J. Res. Natl. Std.*, **50** 201.
Brauer, G.M. (1965), *J. Polymer Sci.*, **8** 3.
Brown, J.E., Tryon, M. & Mandel, J. (1963), *Anal. Chem.*, **35** 2172.
Burns, C.L., Forziati, A.F. & Brauer, G.M. (1957), *Intern. Assoc. Dental Res., 35th meeting, Atlantic City, N.J.*
Challinor, J.M. (1983), *Forensic Science International*, **21** 269.
Chiavari, G., Ferretti, S., Galletti, G.C. & Rocco, M. (1991), *J. Analytical and Applied Pyrolysis*, **20** 253.
Clausen, P.K. & Rowe, W.F. (1980), *J. Forensic Sciences*, **25** 765.
Cleverley, B. & Herrmann, R. (1960) *J. Appl. Chem.*, **10** 192.
Colling, E.L., Burda, B.H. & Kelley, P.A. (1986), *J. Chromatographic Science*, **24** 7
Curry, C.J. (1987), *J. Analytical and Applied Pyrolyses*, **11** 213.
Dandoy, J. (1962), *Ind. Chim. Belge*, **27** 355.
Davidson, W.H.T., Slaney, S. & Wragg, A.L. (1954), *Chemical Industry* 1356.
DeAngelis, G., Ippoliti, P. & Spina, N. (1958), *Ric. Sci.*, **28** 1444.
DeForest, P.R., Doctoral disertation, University of California at Berkeley, June 1969.
DeForest, P.R. (1974), *J. Forensic Sciences*, **19** (1) 113.
Dhont, J.H. (1961), *Nature*, **192** 747.
Dhont, J.H. (1962), *Chem. Weekblad*, **58** 440.
Dhont, J.H. (1963a), *Nature*, **198** 990.

Dhont, J.H. (1963b), *Nature*, **200** 882.
Dhont, J.H. (1964), *Analyst*, **89** 71.
Duncan, W.P. (1988), American Laboratory, **20** 40.
Esposito, G.G. (1964), *Anal. Chem.*, **36** 2183.
Ettre, K. & Varadi, P.F. (1962), *Anal. Chem.*, **34** 752.
Fiegl, F. & Jungreis, E. (1958), *Mikrochem Acta*, 812.
Fjeldsted, J.C. & Lee, M.L. (1984), *Analytical Chemistry*, **56** 619A.
Fontan, C.R., Jain, N.C. & Kirk, P.L. (1964), *Mikrochim Acta*, 362.
Garzo, G. & Szekely, T. (1964), *Acta Chim. Acad. Sci. Hung.*, **41** 269.
Gere, D.R. (1983), *Science*, **222** 253.
Giddings, J.C., Meyers, M.N., McLaren, L. & Keller, R.A. (1968), *Science*, **162** 67.
Glassner, S. & Pierce, R. (1965), *Anal. Chem.*, **37** 525.
Gouw, T.H. & Jentoft, R.E. (1972), J. Chromatography, **68** 303.
Granada, E., Blasco, J., Comellas, L. & Gasiot, M. (1991), *J. Analytical and Applied Pyrolysis*, **19** 193.
Groten, B. (1964), *Analytical Chemistry*, **36** 1206.
Guiochon, G. & Henniker, J. (1964), *Brit. Plastics*, **37** 74.
Happ, G.P. & Maier, D.P. (1964), *Anal. Chem.*, **36** 1678.
Harms, D.L. (1953), *Anal. Chem.*, **25** 1140.
Hawley-Fedder, M.L., Parsons, M.L. & Karasek, F.W. (1984), *J. Chromatography*, **314** 263.
Hurd, C.D. (1929), *The pyrolysis of carbon compounds*, The Chemical Catalog Company, Inc., New York, 9.
4th International Symposium on Supercritical Fluid Chromatography and Extraction, Cincinnatti, Ohio, May 1992.
Jackson, W.P., Markides, K.E. & Lee, M.L. (1986), *J. High Resolution Chromatography and Chromatography Communications*, **9** 213.
James, A.T. & Martin, A.J.P. (1952), *Analyst*, **77**, 915.
Janak, J. (1960a), *Nature*, **185** 684.
Janak, J. (1960b), in *Gas chromatography* R.P.W. Scott (ed.) Butterworths, London, 387.
Jones, C.E.R. & Moyles, A.F. (1961), *Nature*, **189** 222.
Karr, C., Comberiati, J.R. & Estep, P.A. (1961a), *Fuel*, **42** 211.
Karr, C., Camberiati, J.R. & Warmer, W.C. (1963b), *Anal. Chem.*, **35** 1441.
Keulemans A.I.M. (1958), in *Gas chromatography*, A.J.P. Martin (ed.) Academic Press, New York, 237.
Kirk, P.L. (1963), *Science*, **140** 367.
Klesper, E., Corwin, A.H. & Turner, D.A. (1962), *J. Organic Chemistry*, **27** 700.
Kruse, P.F. & Wallace, W.B. (1953), *Anal. Chem.*, **25** 1156.
Kupfer, W. (1962), *Z. Anal. Chem.*, **192** 219.
Lai, S.T. & Locke, D.C. (1983), *J. Chromatography*, **255**, 511.
Lehman, F.A. & Brauer, G.M. (1961), *Anal. Chem.*, **33** 673.
Levy, R.L. (1963), Thesis, Israel Inst. Technology, Haifa, Israel, August.
Levy, R.L. & Gesser, H. (1965), *Proc. Intern. Congr. Chromatog., Athens*, 263.

Levy, R.L., Gesser, H., Halevi, E.A. & Saidman, S. (1964), *J. Gas Chromatography*, **2** 254.
Lloyd, J.B.F., Hadley, K. & Roberts, B.R.G. (1974), *J. Chromatography*, **101** 417.
Martin, S.B. (1959), *J. Chromatography*, **2** 272.
Martire, D.E. & Boehm, R.E. (in press), *J. Physical Chemistry*.
May, R.W., Pearson, E.F. & Scothern, M.D. (1973), *Analyst*, **98** 364.
McMinn, D.G., Carlson, D.G. & Munson, T.O. (1985), *J. Forensic Sciences*, **30(4)** 1064.
Meuzelaar, H.L.C., Windig, W., Harper, A.M., Huff, S.M., McClennen, W.H. & Richards, J.M. (1984), *Science*, **226** 268.
Meyers, M.N. & Giddings, J.C. (1966), *Separation Science* **1** 761.
Munson, T.O. & Vick, J. (1985), *J. Analytical and Applied Pyrolysis*, **8** 493.
Nelson, D.F. & Kirk, P.L. (1962), *Anal. Chem.*, **34** 899.
Newlon, N.A. & Booker, J.L. (1979), *J. Forensic Sciences*, **24** 87.
Norris, T.A. & Rawden, M.G. (1984), *Analytical Chemistry*, **56** 1767.
Novotny, M. (1986), *J. High Resolution Chromatography and Chromatography Communications*, **9** 137.
Oyama, V.I. (1963a), *Nature*, **200** (2), 1058.
Oyama, V.I. (1963b), *Lunar and Planetary Exploration Coll.*, **3** 29.
Parson, J.S. (1964), *Anal. Chem.*, **36** 1849.
Peadon, P.A. & Lee, M.L. (1983), *J. Chromatography*, **259** 1.
Pentoney, S.L., Shafer, K.H. & Griffiths, P.R. (1986), *J. Chromatographic Science*, **24** 230.
Raddell, E.A. & Strutz, H.C. (1959), *Analytical Chemistry*, **31** 1890.
Rothchild, R. & DeForest, P.R. (1982), *J. High Resolution Chromatography and Chromatography Communications*, **1** 321.
Schwartz, H.W. (1987), *LC/GC*, **5** 14.
Silverman, L., Trego, K., Honk, W. & Shideler, M.E. (1958), *J. Appl. Chem.*, **8** 616.
Simon, W. & Giacobbo, H. (1965), *Chem. Ing. Tech.*, **37** 709
Smith, R.D., Kalinosky, H.T., Udseth, H.R. & Wright, B.W. (1984), *Analytical Chemistry* **56** 2476.
Stanley, C.W. & Peterson, W.R. (1962), *Soc. Plastics Engineers Trans.*, **2** 298.
Stewart, W.D. (1974), *J. Forensic Sciences*, **19(1)** 121.
Voigt, J. (1964), *Kunstoffe*, **54** 2.
Voigt, J. & Fischer, W.G. (1964), *Chemiker Ztg.*, **88**, 919.
West, W.R. & Lee, M.L. (1986), *J. High Resolution Chromatography and Chromatography Communications*, **9** 137.
Wheals, B.B. & Noble, W. *J. Forensic Sciences*, **14 (1)** 23.
Wright, M.M. (1987), *J. Analytical and Applied Pyrolysis*, **11** 195.
Zemany, P.D. (1952), *Analytical Chemistry*, **24** 1709.

Index